Coping With Numbers

Coping With Numbers

A Management Guide

DAVID TARGETT

Basil Blackwell • Oxford

First published 1983 by Martin Robertson & Company Ltd.
Reprinted 1984, 1987, 1988, 1990 and 1992 (twice), 1994 by
Blackwell Publishers
108 Cowley Road, Oxford, OX4 1JF, UK

British Library Cataloguing in Publication Data

Targett, David
    Coping with numbers.
    1. Numerical analysis
    I. Title.
    519.4   QA297
    ISBN 0–631–14123–5

Typeset by Styleset Limited - Salisbury
Printed and bound in Great Britain
by T. J. Press (Padstow) Ltd

This book is printed on acid-free paper.

Contents

Preface

There is a gap between what managers are offered to help them cope with the numbers they encounter in their work and what they need. My principal reason for writing this book is to fill this gap, at least in part. Many managers feel confused by all the data that circulate in modern organizations. When they turn to 'numbers experts' (typically statisticians, operational researchers and computer scientists), they do not always receive the help they either want or deserve. The traditional numbers subjects have become increasingly sophisticated; so have the experts. This has two effects. First, the managers find it harder to understand the experts. The quality of communication between managers and experts tends to be low. Second, the experts are accustomed to large, technically complex projects and find it difficult to operate at a simpler level. In any case, the numbers subjects usually taught to managers were not developed for management. Statistics was developed for the sciences and operational research for military uses. These traditional subjects do not always deal with management problems satisfactorily. They do not touch some areas of management at all and where they are useful they are often expounded in a way that is more concerned with mathematical rigour than applicability.

These are sizeable problems and it would be an ambitious book that set out to solve them. 'Coping with Numbers' is not so ambitious, but does attempt to do something in this direction. It suggests some new techniques where there are currently gaps; it describes the underlying concepts and practicalities of the more useful of the traditional techniques. It is in four parts, each of which is virtually self-contained. In Part I are methods for dealing with large sets of numbers. Part II highlights those areas of statistics that can help the way a manager thinks. Time and expense is currently being devoted to business forecasting which is the main topic of Part III. Part IV highlights those parts of operational research that are especially useful in decision-making. Each part has an introduction describing its purpose and rationale. Throughout, there are worked examples which, without exception, are derived from real problems in management. Further examples are given, with answers in outline only. A glossary of specialist terminology is given at the end of the book for easy reference.

With these objectives and such a wide coverage it goes without saying that the level of mathematical rigour is intentionally not high. It is not intended to turn readers into practitioners; it is intended to demonstrate what is practical and useful for managers, giving them guidance in day to day tasks of number handling. In the less technical areas there are techniques for manipulating numbers and converting them into useful information; in the complex areas the emphasis is on the concepts and practical aspects so that the manager can at the very least participate sensibly in discussions with the experts.

The book is therefore for managers. They may be practising managers or managers in training. Those in training may be on courses in business studies or related subjects such as economics and accounting. The technical content of the book will probably be insufficient for the more mathematical courses in statistics and operational research. In these cases the role of the book is in giving a practical orientation to specialist skills.

A word of warning is needed. Even though the book is for non-specialists, it is not necessarily easy. It is not possible to cover these subject areas without becoming involved in a few technicalities. Technical passages are well-marked and the non-numerate should think about missing them out on a first reading. An eminent statistician once said of one of his books that nobody should read it unless they had read it before. The hope is that any struggle will be worthwhile. It is now being more and more recognized that there is a need for a greater sense of analysis in management decisions.

I am indebted to friends and colleagues who have helped me to write the book. My thanks go to Professor David Chambers for his advice and encouragement. Also at London Business School I am grateful to Duncan Baird, John Eaton, Mary Jackson, David Juster and Jacqueline Ord for giving their time to read the early drafts, make invaluable comments and correct mistakes.

Turning Numbers into Information

The subject of statistics has traditionally been regarded as providing ways of analysing with large quantities of numbers. Unfortunately, statistical methods were not developed with management in mind. Nor were they developed in the computer era with its resultant proliferation of data. Whilst statistics is still relevant, at least in part, some additional methods are required to meet current management needs. Many managers claim that they have insufficient information to do their job as they would like. It may be more true to say that they have too many data which are difficult to convert into useful information. The three chapters in Part I attempt to deal with this problem by showing how data can be manipulated and turned into information. They also show how information contained in tables and graphs can be communicated to others.

Preparation

The technical content of the three chapters of Part I is small and there is no prerequisite knowledge.

Turning Numbers into Information

Data communication

By the end of the chapter the reader should be better able to present tables, accounting data and graphs so that the information contained in them is immediately apparent to users. The methods are not just of use in communicating data; they will be of fundamental importance in the analysis of data.

Data communication means the transmission of information through the medium of numbers. Popular opinion would seem to be that it is generally not done well. Either it is done dishonestly ('figures can lie and liars can figure'); or it is done confusingly so that the numbers appear incomprehensible and any real information is obscured. Thus far the situation is the same for numbers as for words. The difference is that numbers are abandoned as a lost cause. More effort is made with words. One hears, for instance, of compaigns for the plain and efficient use of words by bureaucrats, lawmakers, etc., but not for the plain and efficient use of numbers by statisticians, computer scientists and accountants. Furthermore, whilst experts spend much time devising advanced numerical techniques, little effort is put into methods for better data communication.

This chapter attempts to redress the balance in some way by dealing with the question of data communication. Numbers are nearly always produced in the form of tables of varying types, or in the form of graphs and similar pictorial methods. The role of both these modes of presentation will be discussed. The case of accounting data will be given separate treatment. The chapter relies heavily on the work of A.S.C. Ehrenberg referenced in the bibliography.

The aim of the chapter is to show how data might be presented better. In what way is 'better' better? A manager meets data in just a few general situations:

- *Business reports*: the data are usually the supporting evidence for conclusions or suggestions made verbally in the text.
- *Management information systems*: large amounts of data are delivered to the manager at regular intervals.
- *Accounting data*: primarily, for a manager, these will indicate the major financial features of an organization. The financial analyst will have more detailed requirements.
- *Self-generated data*: the manager wishes to analyse his own data.

In all these situations speed is essential. A manager is unlikely to have the time to carry out a detailed analysis of every set of data that crosses his desk. The features of the data should be immediately obvious to him. Moreover the features should be the main ones rather than the points of detail. These requirements suggest that the criterion that distinguishes well presented data should be:

The main patterns and exceptions in the data should be immediately evident.

The achievement of this objective is made easier since in all the situations above the manager will normally be able to anticipate the patterns in the data. In the first case above the pattern will have been described in the text; in the other three it is unlikely that he will be dealing with raw data in a totally new set of circumstances and he will therefore have some idea what to expect. Some rules for improving data presentation are put forward with the criterion in mind.

In looking at this subject it must be stressed that it is not just communicating data to others that is important. Communicating data to oneself as a step in coming to understand them is perhaps the most valuable function of the ideas proposed here. The role of data communication in the analysis of data will be explored in the next chapter.

Rules for data presentation

There are 7 rules for data presentation. They should improve the communication of any information contained in the data. Not all the rules can be applied to any one set of data.

Rule 1   Round the numbers to 2 effective figures

When trying to understand numbers, most people round them mentally to bring them to a size that the mind can manipulate more easily. In practice the rounding is to 2 figures. For instance, someone wanting to make the calculation

$$34.8 \div 18.3$$

would probably round the division to

$$35 \div 18$$

to obtain the answer of just less than 2. Of course, there are many situations, an engineer's structural calculations for example, in which rounding would be wholly inappropriate. For management information, rounding will not usually affect decisions being taken, but it will improve communication. If simple numerical patterns exist in the data they will be more readily seen as a result of rounding, since the numbers can now be assimilated mentally.

Rounding to 2 effective figures is variable rounding. Not all numbers will be rounded in the same way. Here are some examples of rounding to 2 effective figures.

Original	*Rounded*
1382	1400
721	720
79.311	79
17.1	17
4.2	4.2
2.32	2.3

These numbers have been rounded to the first 2 figures. Contrast this to fixed rounding, for example always to the first decimal place.

The situation is slightly different if a series of similar numbers all of which have, say, the first 2 figures in common are being compared.

The rounding would then be to the first 2 figures that are effective in making the comparison; i.e. the first 2 that differ from number to number. This is the meaning of 'effective' in '2 effective figures'. For example, the column of numbers below would be rounded as shown:

Original	Rounded
1142	1140
1327	1330
1489	1490
1231	1230
1588	1590

The numbers have been rounded to the second and third figures because all the numbers have the first figure (1) in common. Rounding to the first 2 figures would be over-rounding, making comparisons too approximate.

Many managers may be concerned that rounding leads to inaccuracy. It is true, of course, that rounding does lose accuracy. The important questions are: 'Would the presence of extra figures effect the decision being taken' and 'Just how accurate are the data anyway – is the accuracy being lost spurious accuracy?' Often one finds that 8-figure data are being insisted upon in a situation where the decision being taken rests only on the first figure and where the method of data collection was such that only the first figure can be relied upon. In cases of doubt A.S.C. Ehrenberg suggests that the numbers be rounded but that a note at the bottom of the table indicate a source from which data of greater precision can be obtained. Then wait for the rush.

Rule 2   Reorder the numbers

The pattern in a set of numbers can be understood better when they are in size order. A table of data will then appear more orderly and relationships between the numbers will often be highlighted. For instance, in a financial table, listing the divisions of a company in size order (capital employed, perhaps) shows at a glance whether turnover, profit, etc. are in the same order. Divisions that are out of line in this respect will be pinpointed.

The quantity used to do the ordering may be (a) one of the columns or rows of the table; (b) the averages of the columns or rows; (c) something external to the table, such as ordering geographical regions by population. Frequently tables are ordered alphabetically.

This is helpful in a long reference table which is unfamiliar to the user, but not so helpful when management information is involved. Indeed, it may be a hindrance. In management information it is the overall pattern, not the individual entries, that is of interest. Alphabetical order is more likely to obscure rather than highlight the pattern. In addition, managers are usually not totally unfamiliar with the data they receive. For instance, anyone looking at some product's sales figures by state in the USA would probably be aware that, for a table in population order, California would be close to the top and Alaska close to the bottom. In other words the loss from not using alphabetical order is small, whereas the gain in data communication is large.

Rule 3   *Interchange rows and columns*

Making a comparison between numbers or seeing a pattern in them is easier if the numbers lie underneath rather than alongside one another. The reason is that it is quicker to see the difference between, say, 2- and 3-figure numbers and to see the size of the difference by subtraction if the numbers are in a column not a row. This probably originates from the way children in school are taught to manipulate numbers. People learn to do sums in this form:

$$
\begin{array}{r}
57 \\
-23 \\
\hline
34
\end{array}
$$

In a table, then, the more important comparison should be down columns not along rows.

Rule 4   *Use summary measures*

Summary measures, nearly always averages, for rows and columns, provide a focus for the eye as it moves down or along the numbers. It is also a basis for comparison, making it easier to tell at a glance the degree of variability and to discern whether a particular element falls above or below the general level of the rest of the column.

If the summary measure is not an average it is important that it should be of the same order of size as the column. A column total is not a summary and is of a different order of magnitude than the rest of the numbers. The summary measure can also be the basis for ordering the rows and columns (see rule 2 above).

Rule 5   *Minimize the use of space and lines*

There should be as little white space, and as few grid lines, as possible in a table of numbers. Lots of white space may look attractive, but the resultant gaps between the numbers reduce the ability to see patterns, simply because there is a longer interval as the eye travels from one number to the next. The numbers should be close together, but not, of course, so close that they become confused. Likewise grid lines (horizontal or vertical) in a table interfere with the eye's movement from one number to the next. Grid lines should be used to separate one type of number from another; e.g. to separate a summary row from the rest of the numbers. Grid lines should not be used to separate like numbers when the purpose of the table is to compare them.

Rule 6   *Labelling should be clear but unobtrusive*

Care should be taken when labelling data otherwise the labels may confuse and detract from the numbers. The constructor of a table is familiar with the numbers and their definition. He may use abbreviated or obscure labels, falsely assuming that the reader has the same familiarity. Labels should be clear and not interfere with the understanding of the numbers. This means that a gap in a column of numbers would not be introduced merely to accommodate an extra-long label.

Rule 7   *Use a verbal summary*

A verbal summary can help achieve the objective of quickly communicating information by directing attention to the main features of the data. The summary should be short and deal with the main pattern, not the points of detail nor minor eccentricities in the data. It should not, of course, mislead.

In a management report, the verbal summary will probably be there already since the data are likely to be present to provide evidence of a conclusion drawn and described in the text. In other circumstances where a manager meets data, a verbal summary should be produced, if possible, along with the data.

EXAMPLE: INTERNATIONAL FACTS

Table 1.1 is taken from a paperback book, which is full of tables of economic and other statistics compiled by the Central Statistical

Table 1.1   Gross domestic product at current market prices
(Eur. thousand million)

	1965	1970	1972	1973	1974	1975
UK	99.3	121.7	143.6	140.4	152.8	172.5
Belgium	16.6	25.2	31.6	35.9	42.0	46.2
Denmark	10.1	15.6	19.2	21.8	24.3	26.9
France	95.8	140.9	176.6	200.5	212.6	253.3
Germany (FR)	114.3	185.5	235.7	275.4	305.8	319.9
Irish Republic	2.7	3.9	5.1	5.2	5.4	5.9
Italy	58.4	92.7	109.4	112.7	122.1	130.2
Luxembourg	0.7	1.1	1.2	1.5	1.7	1.7
Netherlands	18.7	31.6	41.7	48.4	55.8	61.2
Japan	89.0	196.9	270.9	327.5	365.6	372.8
USA	690.0	984.4	1,076.0	1,041.2	1,125.8	1,149.7

Office of the United Kingdom government. The table shows the Gross Domestic Product of the then nine countries of the European Economic Community for various years between 1965 and 1975 with the USA and Japan included as a comparison. It is intended as background information only. An economist doing a detailed study of any or all of the countries would refer to sources other than a general statistical paperback book. Typically a reader may wish to ask questions such as: How does the economic growth of West Germany compare with the UK? How do the economic sizes of the major EEC countries compare one with another? Is Japan catching up the USA? Just how small are the smaller EEC countries? The table is presented neatly, but typical questions such as the above take a surprisingly long time to answer. A few changes to the layout along the lines of the 7 rules make the information much easier to grasp.

After applying the 7 rules the amended data are shown in table 1.2. The rules were applied as follows:

Rule 1   Rounding to two effective figures is straightforward except for the largest country, the USA, and the two smallest, the Irish Republic and Luxembourg. The data for the USA are rounded to the nearest 10 even when over 1000 since the only numbers above 1000 all have 1 as the first figure. For the two smallest countries the data have been over-rounded. Two effective figures should mean that the decimal place is retained. To do so, however, would upset the balance

Table 1.2   Gross domestic product at current market prices (amended)
(Eur. thousand million)

	1975	1974	1973	1972	1970	1965
USA	1150	1130	1040	1080	980	690
Japan	370	370	330	270	200	90
Germany (FR)	320	310	280	240	190	110
France	250	210	200	180	140	100
UK	170	150	140	140	120	100
Italy	130	120	110	110	93	58
Netherlands	61	56	48	42	32	19
Belgium	46	42	36	32	25	17
Denmark	27	24	22	19	16	10
Irish Republic	6	5	5	5	4	3
Luxembourg	2	2	1	1	1	1

of the table and attract over-much attention to these countries. The loss of accuracy is not serious, because comparisons involving them can still be made. In any case students of the economic progress of Luxembourg would doubtless not be using this book as a source of data.

Rule 2   The rows are ordered by size. The previous order was UK first (it is a UK publication) followed by the other EEC countries alphabetically, followed by the two non-EEC countries. Alphabetical order does not help in answering the questions, nor does the separation of the USA and Japan (does anyone doubt that Japan is not an EEC member?). Size order makes it easier to compare countries of similar size as well as immediately indicating changes in ranking. The ordering in table 1.2 is based on the 1975 GDP. It could equally have been based on 1965 GDP or the populations. There is often no one correct method of ordering. It is a matter of taste. The columns are ordered from 1975 to 1965 so that the most recent data are next to the labels. This is also a matter of taste. Some people prefer chronological ordering, moving left to right from 1965 to 1975.

Rule 3   Interchanging rows and columns is not necessary. One is just as likely to wish to compare countries (down the columns) as one country across the years (along the rows).

Rule 4   The use of summary measures is not particularly helpful.

Row and column averages would have no intuitive meaning. For example, the average GDP for a country over a selection of years is not a useful summary.

Rule 5 The vertical lines in the original table hinder comparisons across the years and have been omitted.

Rule 6 The labelling is already clear. No changes have been made.

Rule 7 No simple verbal summary is possible for these data. Moreover, the publishers would probably not wish to be appearing to lead the reader's thinking by suggesting what the patterns were.

The typical questions that might be asked of these data can now be applied to Table 1.2. It is possible to see quickly that West Germany's GDP increased by 320/110 = just under 3 times; Italy's by 130/58 = just over 2 times; Japan's by just over 4 times; the UK's by 1.7; Japan has overtaken West Germany, France and the UK; the Irish Republic is 3 times the size of Luxembourg economically. The information is more readily apparent from Table 1.2 than Table 1.1.

The special case of accounting data

The communication of accounting data requires special treatment. The main factors that make accounting data different are:

- they have a logical sequence e.g. a Revenue and Expenditure statement builds up to a final profit;
- there is a tradition of how accounts should be presented;
- there are legal requirements.

These factors do not mean that the presentation of accounts cannot be improved, but they do mean that proposed changes will be met with resistance. The difficulty seems to be that what constitutes a well-presented set of accounts to an accountant may differ from what constitutes a well-presented set of accounts to a layman. The accountant's criteria are probably based on what is good practice when drawing up accounts and on whether other accountants approve of them. For example most final accounts use brackets to indicate a negative. This makes sense when preparing accounts and when dealing with other accountants, but most laymen are more used to a minus sign to indicate a negative. Yet many financial data are for laymen. Published accounts are for shareholders; internal company information is for managers (not all of whom are former accountants).

The accounts in table 1.3 are a case in point. They are an extract

Table 1.3   *Statement of revenue and expenditures*

		1981	1980
		£	1000 £
Gross Freight Income		98 898 684	73 884
Voyage Expenses:			
Own Vessels	31 559 336		(24 493)
Vessels on timecharter	42 142 838	(73 702 174)	(28 378)
Operating Expenses:			
Crew Wages and Social Security	7 685 965		(9 010)
Other Crew Expenses	541,014		(633)
Insurance Premiums	1 161 943		(1 367)
Provisions and Stores	1 693 916		(2 268)
Repairs and Maintenance	1 685 711		(3 297)
Other Operating Expenses	60 835		(27)
		(12 829 384)	
Result from Vessels		12 367 126	4 411
Results from Parts in Limited Partnerships		1 793 314	(163)
Result from Operation Business Building		167 343	0
Management Expenses			
Salaries, Fees, Employees' benefits etc	1 426 607		(1 208)
Other Management Expenses	502 815		(635)
		(1 929 422)	
Depreciation Fixed Assets		(7 106 305)	(5 365)
Transferred from Classification Survey Fund	112 401		652
Set aside for Classification Survey Fund	(612 401)		(27)
		(500 000)	
		4 792 056	(2 335)
Capital Income and Expenses:			
Misc. Interest Income	318 601		364
Misc. Interest Expenses	(8 450 307)		(6 482)
	(8 131 706)		
Net Profit/(Loss) currency exch	(190 836)		(680)
Dividends	35 732		47
		(8 286 810)	
Other Income and Expenses			
Net Profit from sales Fixed Assets	5 553 047		5 593
Balance Classification Survey and Self Insurance Fund upon Sale	1 414 132		858
	6 967 179		
Net Profit/(Loss) from sale of shares	(38 647)		0
Misc. Income/(Expenses) relating to previous years	629 630		1 408
Adjustm. cost shares	226 197		(206)
Loss from Receivables	(219 189)		0
Reversal write up Fixed Assets previous years	(2 026 067)		0
		5 539 103	
Result before Taxes and Allocations		2 044 349	(1 433)
Allocations		(1 929 400)	
		114 949	(1 433)
Reserves	(11 495)		1 071
Dispositions	(103 454)		362
		(114 949)	1 433

from the accounts of a shipping company. An accounting association awarded them a prize for their excellence as examples of annual accounts. Whilst they doubtless represent fine accounting practice, it is almost impossible to tell what, financially, befell the company between the 2 years. If an attempt is made to improve the communication of the data by applying some of the 'rules' (only rules 1, 5, 6 and 7 are appropriate for accounts), some interesting facts emerge. Table 1.4 shows the same accounts re-presented. The new table reveals that it was an unusual year for the company. Incomes and expenditures changed by large and extremely variable amounts.

Freight income	up 33%
Voyage expenses	up 40%
Operating expenses	down 20%
Operating profit	up 180%
Partnership shares	small negative to large positive
Management expenses	equal
Capital income expense	up 25%
Other income expense	down 25%
Profit before tax	negative to positive

Table 1.4   Statement of revenue and expenses (amended)

	Million £	
	1981	*1980*
Gross freight income	98.9	73.9
Voyage expenses	−73.7	−52.9
Operating expenses	−12.8	−16.6
PROFIT FROM VESSELS	12.4	4.4
Shares in partnership	1.8	−0.2
Operation business building	0.1	0
Management expenses	−1.9	−1.8
Depreciation	−7.1	−5.4
Classification survey (net)	−0.5	0.6
Capital income and expenses	−8.3	−6.7
Other income and expenses	5.5	7.7
PROFIT BEFORE TAX	2.0	−1.4
Allocations	−1.9	0
TO RESERVES	0.1	−1.4

These are the main features of the company's finances. Even knowing what they are it is very difficult to see them in the original table. Yet it is this volatility that is of major interest to shareholders and managers.

Communicating financial data is an especially challenging area. The guiding principle is that the main features should be evident to the users of the data. It should not be necessary to be an expert in the field nor to have to carry out a complex analysis in order to appreciate the prime events in a company's financial year. Some organizations are recognizing these problems by publishing two sets of (entirely consistent) final accounts. One is source material, covering legal requirements and suitable for financial experts; the other is a communicating document, fulfilling the purpose of accounts which is to provide essential financial information to shareholders and other non-accountants.

Communicating data through graphs

The belief that a picture speaks a thousand words (or numbers) is widely held. For most people a picture is more interesting and attractive than numbers. A graph is a form of picture which can, in certain circumstances, be very helpful in communicating data. Graphs are not, however, always helpful. It is important to distinguish between the correct and incorrect uses of a graph. Figure 1.1 is a graph of the annual inflation rate in the UK in the years 1976–79. It works well: the pattern can be seen quickly and easily. This is no longer so if the graph is used for comparing inflation in several countries. Figure 1.2 is figure 1.1 with several more countries added. The graph ceases to function well. There are too many lines and they tend to cross over. As a result no overall message emerges. Furthermore the graph cannot be used for reference purposes without difficulty and inaccuracy. What, for instance, was the inflation rate in Belgium in 1977?

This example illustrates the principles underlying the use of graphs for data communication.

- Graphs are good when:
  attracting attention and adding variety to a long series of tables;
  communicating very simple patterns.

- Graphs are not good when:
  communicating even slightly complex patterns;
  being used for reference purposes.

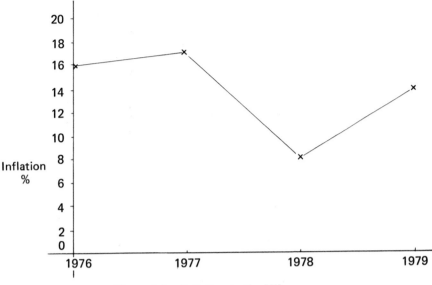

Figure 1.1   Inflation in the UK

These principles are straightforward and it may appear at first sight that little could go wrong in the use of graphs in organizations. However things do go very wrong with surprising frequency. Figure 1.3 is a disguised version of part of the management information system of a large company. It was one of a series of about 40 similar graphs. It is not difficult to see that it has been produced automatically from a computer with little thought to communication (the vertical scale is an indication that it comes hot from the computer). Nor is it difficult to say what is wrong. No pattern of any sort is remotely evident. To extract an individual number (say, the imports for Denmark in March 1977) requires firstly being able to find the line for the country and, secondly, to estimate the imports by a calculation of the type: about two-fifths of the way between 456 and 717....
Since the graph is unusable, at the very least expense could be saved by not issuing it.

These data are difficult to communicate because they are so many. Distinct improvements can, nevertheless, be made, using the rules of presentation.

Table 1.5 gives the same data that are on the graph but in tabular form. It can now certainly be used for reference purposes. In addition, it is possible, although not easy, to see the seasonal pattern in the data.

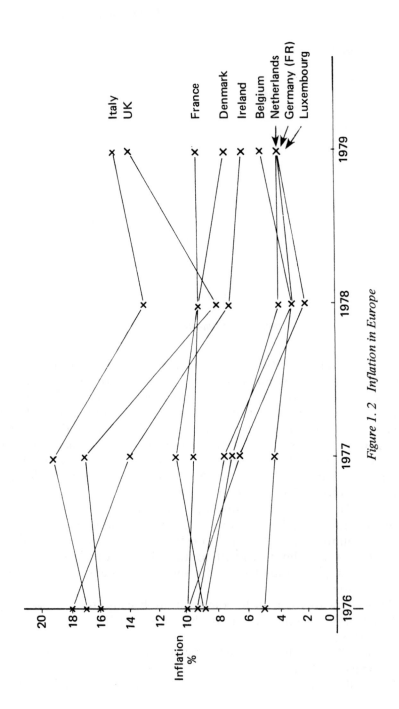

Figure 1. 2 Inflation in Europe

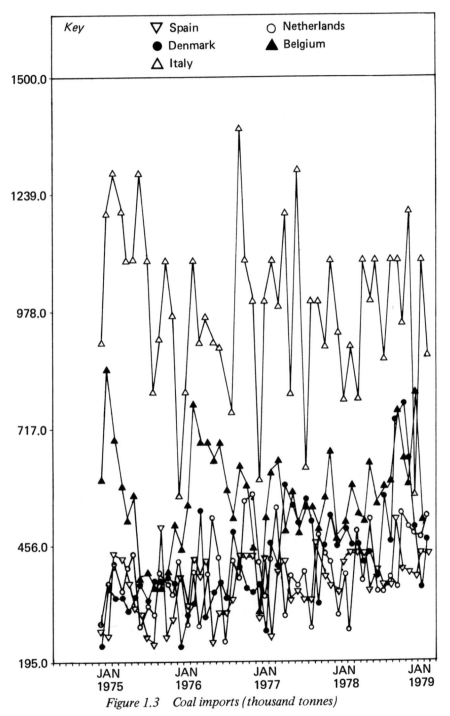

Figure 1.3   Coal imports (thousand tonnes)

Table 1.5   Monthly coal imports (thousand tonnes)

		Italy	Belg.	Den.	Neth.	Spain
1975	Jan	910	610	230	280	260
	Feb	1200	850	360	360	240
	Mar	1300	720	340	420	440
	Apr	1200	590	340	350	430
	May	1100	520	310	410	360
	June	1100	570	330	440	320
	July	1300	380	380	240	290
	Aug	1100	390	330	320	250
	Sept	800	360	370	300	210
	Oct	920	360	380	390	500
	Nov	1100	390	400	370	240
	Dec	970	500	370	340	290
1976	Jan	570	450	230	420	380
	Feb	800	550	300	280	310
	Mar	1100	770	330	400	430
	Apr	910	690	540	270	380
	May	970	690	290	390	420
	June	910	660	350	520	240
	July	900	690	370	430	300
	Aug	900	580	330	240	300
	Sept	750	520	480	430	340
	Oct	1400	640	410	380	430
	Nov	1100	590	360	560	430
	Dec	1000	450	340	570	430
1977	Jan	620	310	360	420	290
	Feb	1000	520	250	320	440
	Mar	1100	630	460	430	260
	Apr	990	660	430	540	410
	May	1200	480	590	290	420
	June	790	580	560	390	330
	July	1300	470	520	370	370
	Aug	640	540	570	400	330
	Sept	1000	540	510	270	330
	Oct	1000	490	320	500	490
	Nov	900	570	470	440	380
	Dec	1100	680	530	420	370

Table 1.5 (continued)

		Italy	*Belg.*	*Den.*	*Neth.*	*Spain*
1978	Jan	930	470	470	300	360
	Feb	780	510	530	390	420
	Mar	900	590	440	260	440
	Apr	780	530	440	490	440
	May	1100	510	420	400	440
	June	1000	650	440	510	350
	July	1100	550	390	350	400
	Aug	870	580	570	350	360
	Sept	1100	610	460	380	360
	Oct	1100	750	730	360	530
	Nov	950	660	750	530	410
	Dec	1200	600	650	500	400
1979	Jan	570	790	490	500	390
	Feb	1100	520	360	470	450
	Mar	880	870	460	520	450
Average		990	570	420	400	370

If the general pattern over the years, or a comparison between countries, is required, table 1.6 is suitable. This shows the average monthly imports of coal for each year. It can now be seen that four of the countries have increased their imports by a factor of 30–45%. Italy is the exception, having decreased coal imports by 20%. The level of imports in the countries can be compared. In these terms, Italy is the largest, followed by Belgium, followed by the other three at approximately the same level.

Table 1.6 can be transferred to a graph as shown in figure 1.4. General patterns are evident. Italy has decreased its imports, the others increased; the level of imports is in the order Italy, Belgium . . . The difference between the table and the graph is in estimating magnitudes. The percentage change (−20% for Italy, etc.) is readily calculated from the table, but not from the graph. The purpose of the data and personal preference would dictate which of the two were used.

Graphs are the most important, but not the only, pictorial method

Table 1.6   Annual coal imports as monthly averages (thousand tonnes)

	Italy	Belg.	Den.	Neth.	Spain
1975	1100	520	340	350	320
1976	940	610	360	410	370
1977	970	540	460	400	370
1978	980	580	520	400	410
1979	850	730	440	500	430
Average	990	570	420	400	370

of communicating numbers. A wide range of possibilities exist, all of which have their advantages and disadvantages. The underlying principles are the same as for graphs. Pictures are useful for attracting attention and for showing very general patterns. They are not useful for showing complex patterns or for extracting actual numbers.

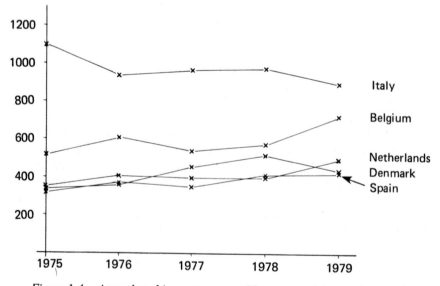

Figure 1.4   Annual coal imports as monthly averages (thousand tonnes)

Worked examples

1. BRITISH ROAD USERS MAGAZINE

Table 1.7 is taken from a major monthly publication on sale to the general public in the UK. It shows the market share for new car sales in the UK for major producers in a particular month. No further explanation was given in the magazine but the figures in parentheses refer to the previous month. Given that the table is background information for a general readership, how could it be

Table 1.7   Market shares for new car sales

British cars	percent
BL Cars	29.31 (25.22)
(includes 7.8% imported)	
Vauxhall	10.47 (8.45)
(includes 11.9% imported)	
Chrysler	10.23 (8.05)
(includes 31.9% imported)	
Ford	5.82 (18.72)
(includes 45.8% imported)	
Rolls-Royce/Bentley	0.10 (0.10)
Reliant	0.09 (0.08)
Lotus	0.08 (0.06)
Panther	0.02 (0.01)

Imported cars	percent
Datsun	6.08 (6.30)
Fiat	5.29 (4.78)
Renault	5.22 (4.94)
VW/Audi	5.19 (3.85)
Chrysler	3.3 (1.9)
Peugeot	3.00 (2.16)
Ford	2.7 (8.03)
Volvo	2.7 (2.12)
Citroën	2.47 (2.57)
BL Cars	2.3 (1.65)
Opel	2.01 (1.9)
Lada	1.89 (1.18)
Toyota	1.42 (1.74)
Alfa-Romeo	1.33 (1.15)

presented better? The data have been presented poorly and there are many inconsistencies (3.3% appears as 3.3%, 3% appears as 3.00%), ambiguities (does the 7.8% underneath British Leyland mean 7.8% of the market share of 29.31%, or does it mean 7.8% in its own right) and double-counting (BL Cars and Ford appear in both sections) in the table. Using the rules of data presentation the table has been amended as in table 1.8.

Table 1.8   Market shares for new car sales
(amended)

	Nov	Oct
*British cars**		
BL Cars	27	23
Vauxhall	9	7
Chrysler	7	6
Ford	3	11
Others	1	1
Imported cars		
Datsun	6	6
Fiat	5	5
Renault	5	5
VW/Audi	5	4
Chrysler	3	2
Peugeot	3	2
Ford	3	8
Volvo	3	2
Citroën	2	3
BL Cars	2	2
Opel	2	2
Lada	2	1
Toyota	1	2
Alfa Romeo	1	1
Others	10	7
Totals		
British made	47	48
Imported	53	52

* Only home-produced cars included. British manufacturers import some cars and these are shown under 'Imported cars'.

Rule 1 Rounding to two effective figures would result in one decimal place for each number. It would be perfectly acceptable to do so. Given the general readership of the data, not to mention the likely accuracy of the original data, it is possible to go further and eliminate even the one decimal place.

Rule 2 The data are already in size order.

Rule 3 Rows and columns have not been interchanged. Comparisons between producers are at least as important as between months and, in any case, an interchange would cause problems with the labelling.

Rule 4 A summary measure in the form of an average market share would not be useful since it would depend upon the number of producers included. Note that many producers are already excluded and the cut-off point in terms of market share was probably arbitrary. Subtotals for British and imported cars have been included, not as summary measures, but because this is interesting and newsworthy information.

Rule 5 Labelling can be improved by eliminating the bracketed phrases underneath the British producers who import some cars, e.g. 'includes 7.8% imported'. This long labelling takes attention away from the numbers. Since these cars are already included under the imported cars heading, it is only necessary to mention this fact in a note at the bottom of the table.

Rule 6 The width of the table is governed by printing restrictions on column width and space available. The original table does not take up an unnecessary amount of room, nor does it use grid lines.

Rule 7 Since the table was located in the middle of text, one would expect any verbal summary to be given there. Strangely in this case the text carried no comment on the data. A verbal summary might be: 'November was similar to October apart from a national strike at Ford causing them to lose market share which was taken up by other British producers.'

2. WHITE AND SNOW HOLDINGS

Table 1.9 is taken from the annual accounts of the company. A second Profit and Loss statement later in the accounts is more detailed and presumably satisfies legal requirements and the needs

Table 1.9   Profit and loss for White and Snow Holdings: summary of combined figures

£ million	1981	1982
Results for the year ended 31 December		
Sales to third parties	4573	4922
Operating profit	272	294
Concern share of associated companies' profit	29	32
Non-recurring and financial items	25	29
Profit before taxation	276	297
Taxation	136	148
Profit after taxation	140	149
Outside interests and preference dividends	20	21
Profit attributable to ordinary capital	120	128
Ordinary dividends	48	52
Profit of the year retained	72	76

of financial specialists. Table 1.9 is therefore intended to communicate the main features of the company's financial year to non-specialists who may be shareholders, trades unionists, etc. Compared with many accounting statements table 1.9 is already well presented, but what further improvements might be made? Several changes have been made, as shown in table 1.10. The changes are:

1   A small amount of further rounding has been carried out. The 2 effective figures rule has not been adhered to in every case. The reason is that rounding and exact adding up are not always consistent. One has to decide which is the more important – the better communication of the data or the need to allow readers to check the arithmetic. The balance of arguments must weigh in favour of the rounding. Checking totals is a trivial matter at this stage (although not of course in the process of auditing). If a mistake were found in the published accounts of such a large company the fault would almost certainly lie with a printer's error, but 'total checking' is an obsessive pastime and few companies would risk the barrage of correspondence that would undoubtedly ensue even though a note to the accounts explained that rounding was the cause. Because of this factor the two effective figures rule has been broken so that adding and subtracting

is exact. The only remaining question is to wonder why totals *are* exactly right in company accounts which have usually been rounded to some degree (the nearest £million in table 1.9). The answer is that figures have been 'fudged' to make it so.

2   Labels have been shortened and brought nearer the numbers.

3   Important sub-totals, such as 'Profit before taxation' and 'Profit after taxation', have been separated and highlighted.

4   The italics, used to denote negatives, have been substituted with minus signs. Tests on the readability of these accounts showed that the italics caused confusion. The accountant's brackets for negatives would probably have been preferable to the italics.

Table 1.10   Profit and loss for White and Snow Holdings Ltd: summary of combined figures (amended). Results for year ending 31 December (£ million)

	1981	*1982*
Sales to third parties	4570	4920
OPERATING PROFIT	272	294
Share associated companies profit	29	32
Non-recurring items	−25	−29
PROFIT BEFORE TAXATION	276	297
Taxation	−136	−148
PROFIT AFTER TAXATION	140	149
Outside interests/preference dividend	−20	−21
Ordinary dividends	−48	−52
RETAINED PROFIT	72	76

3. INTERNATIONAL FACTS

This example refers to tables 1.1 and 1.2 earlier in the chapter. Table 1.1 was the original and table 1.2 an amended version. Could the data have been presented better by means of a graph?

In considering which is the better method of communication, the salient points are:

1   The sizes of the countries are so different that it is difficult to

incorporate them all on the same graph. Figure 1.5 shows the data for all countries except the USA. Figure 1.6 includes the USA. It is always a problem to deal with situations like this without having to resort to artificial devices such as a logarithmic scale or a break in the vertical axis. Both devices are confusing. Figure 1.6 does not work.
2   Just as there is a difficulty with the largest country because of the scale, so there are problems with several smaller countries. It is impossible to make any real comparisons between the five smallest countries.
3   Differences can be seen from graphs, but not the magnitudes of the differences. For example, it can be seen that West Germany has grown more than the UK, but it is not possible to see that the growth factor for West Germany was about 3 and for the UK 1.7.
4   The lines do not cross over and a common problem with graphs is not present here. The major patterns are distinctly visible.

The decision between table 1.2 and figure 1.5 rests on personal choice and circumstances. Table 1.2 is more precise, but figure 1.5 would seem (to some) less boring than a table of numbers.

Final comments

This chapter has attempted to give some guidance on the communication of data. This is an area that has been neglected, presumably because it is technically simple and there is a tendency in quantitative areas to believe that only the complex can be useful. Yet in modern organizations there can be few things more in need of improvement than data communication.

Although the area is technically simple, this does not mean that there are not immense difficulties. What exactly is the readership for a set of data? How can one overcome the common insistence on data specified to a level of accuracy that is not needed by the decision-maker and is not merited by the collection methods? How much accounting convention should be retained in communicating financial information to the layman? What should be done about the aspects of data presentation that are a matter of taste? The guiding principle in these problems is that the data should be communicated according to the needs of the receiver rather than the producer. Furthermore they should be communicated so that the main features

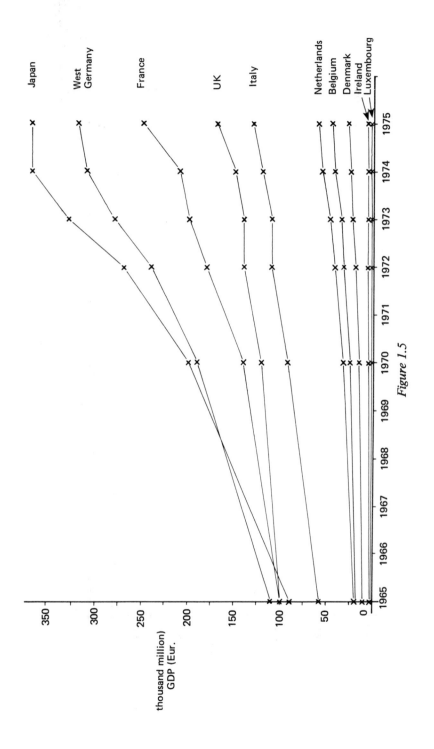

Figure 1.5

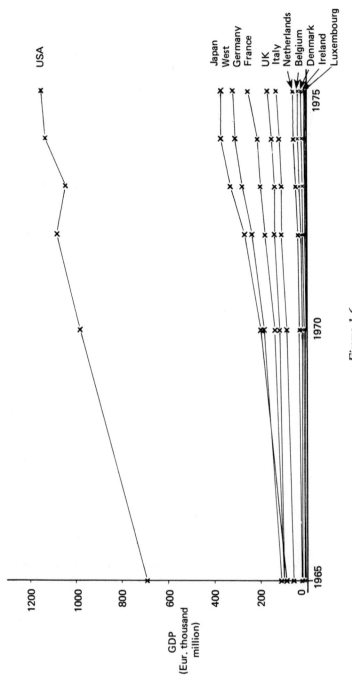

Figure 1.6

can be seen quickly. The seven rules of data presentation described in the chapter seek to accomplish this.

(1) Round to 2 effective figures.
(2) Reorder the numbers.
(3) Interchange rows and columns.
(4) Use summary measures.
(5) Minimize use of space and lines.
(6) Clarify labelling.
(7) Use a verbal summary.

This issue is of especial importance currently because of the growth of microcomputers and the development of large-scale management information systems. The benefits should be enormous but the potential has yet to be realized. The quantities of data that circulate in many organizations are vast. It is supposed that the data provide information which in turn leads to better decision-making. Sadly this is frequently not the case. The data circulate, not providing enlightenment, but causing at best indifference and at worst tidal waves of confusion. Poor data communication is a root cause of this. It could be improved. If not, then one must question the wisdom of the large expenditures many organizations make in providing unused and even bewildering management data.

Further examples

1. OIL AND POWER EFFICIENCY CONSULTANTS

The consultants advise companies how they might save money by making more efficient use of oil. Table 1.11 is part of a recommendation prepared for a client showing the financial return from a capital investment in oil conservation. The investment is evaluated by estimating the internal rate of return (IRR) on the project. Some of the inputs to the calculation, e.g. the life of the project, the level of initial investment, are uncertain. Therefore some sensitivity analysis has been done. The IRR is calculated several times. Each time one of the assumptions is varied, all others remaining fixed at their original values.

Table 1.11 gives the results of the sensitivity analyses. It shows the extent to which the assumptions have been varied and the new IRR

Table 1.11   Sensitivity analysis for oil conservation investment

BASE IRR = 24.93%

	Variation in parameter		New IRR	Difference in IRR
RE-BUILDING INVESTMENT COST (£3.842M)	+£0.5M	(£4.342M)	22.71%	−2.22%
	−£0.5M	(£3.342M)	30.49%	+5.56%
OUTSTANDING LIFE (13 years)	−1 year	(12 years)	23.10%	−1.83%
	−2 years	(11 years)	22.62%	−2.31%
	−3 years	(10 years)	22.06%	−2.87%
	−4 years	(9 years)	21.27%	−3.66%
FUEL CONSUMPTION DIFFERENTIAL (43 TONNES/DAY)	+5T/day	(48T/day)	27.73%	+2.80%
	−5T/day	(38T/day)	21.96%	−2.97%
FUEL PRICE ESCALATION RATE (12%/year)	+1% year	(13%/year)	26.42%	+1.49%
	−1%/year	(11%/year)	23.51%	−1.42%

for each variation. The 'base rate' is the IRR for the original calculation. The table is part of the report to be sent to senior managers of
the client company. How could it be better presented? (Note that it
is not necessary to understand fully the situation in order to propose
improvements to the data communication.)

2. BANK OF SOUTHERN SCOTLAND

Table 1.12 shows the Balance Sheet for the bank taken from its
annual accounts. How might it be presented better?

Table 1.12   Balance Sheet for the Bank of Southern Scotland (£'000)

	31.12.81	31.12.80
Cash and Bank Balances	797921	618208
Deposits	1440041	1391144
Investments	2962303	2720673
Advances	93525	25308
INVESTED FUNDS	5293790	4755333
Tax Recoverable	38208	26621
Interest Receivable	71388	57278
Debtors	9529	12942
	119125	96841
Loans	13317	11021
Premises and Equipment	48123	44072
	61440	55093
TOTAL ASSETS	5474355	4907267
Customers Balances	4806784	4317106
Interest Payable	9694	9298
Creditors	30312	30804
Loans	9391	11791
	4856181	4368999
NET ASSETS	618174	538268
Deferred Tax	46019	38657
Reserves	572155	499611
RESERVES	618174	538268

Data analysis

By the end of the chapter the reader should be able to tackle the problem of analysing data systematically. The major difficulties in understanding management data are not technical. The chapter describes, at a non-technical level, a process for understanding the patterns and regularities in a set of data.

Give a group of managers a table of numbers and ask them to analyse it. Observe the activity that follows. While the group may think they are analysing the numbers, almost certainly they will not be doing so. Most probably they will be 'number picking'. Individual numbers from somewhere in the table which look interesting or which support a long-held prejudice will be selected for discussion. If the data are sales figures, remarks may be made such as: 'I see Western region turned £1.2 million last month. I always said that Jim could do it.' To know whether Jim really can do it requires a different approach. His figures should be put in the context of patterns contained in the whole table. It is necessary to analyse the entire table before sensible remarks about individual numbers can be made. The truth in a quotation from Andrew Lang is apparent.

He uses statistics as a drunken man uses lamp posts – for support rather than illumination.

The purpose of this chapter is to give some guidelines showing how illumination may be derived from numbers. The guidelines give five steps to follow in order to find what real information, if any, a set of numbers contains. They are intended to help a manager understand the numbers he or she has to deal with. A secondary objective is to make the point that in most cases a set of numbers can be analysed satisfactorily without recourse to any sophisticated techniques.

One might have thought that understanding numbers is what the whole subject of statistics is about, and so it is. But statistics was not developed for use in management. It was developed in other fields such as the natural sciences. When transferred to management, there are gaps between what is needed and what statistics can offer. Certainly many managers, having attended courses or read books on statistics, feel that something is missing and that the root of their problem has not been tackled. This and other difficulties involved in data analysis in management will be discussed in the next section. This will be followed by some examples of the types of data analysis a manager might have to undertake. Finally the guidelines intended to help fill this statistical gap will be described and illustrated.

Management problems in data analysis

Understanding numbers is not an easy task. A manager faces a unique set of problems in trying to do so. The nature of these difficulties is not always appreciated. This may partly explain the poor or non-existent data analysis seen (or not seen) in many companies. Rich sources of information, of great value in decision-making, are often ignored. An acceptance that these problems exist is the first step in making an improvement. The set of problems includes the following.

1 The statistical gap

The subject of statistics does not provide all the techniques and methods that a manager would like to have at his disposal. Much of statistics is concerned with the *rigorous* testing of theories. The manager faced with last month's sales figures wants to find the real information in the data. Having found by whatever means that, say, sales have increased everywhere except in the Eastern region, the manager is unlikely to be concerned with the rigorous statistical testing of this information. He will test its validity by a wider range of methods including qualitative ones. He would only rarely require a high degree of statistical rigour. Contrast this with the natural sciences where the emphasis is on thoroughly testing under different conditions theories often derived by non-statistical methods (such as being hit on the head by an apple?). What the manager primarily needs, and what statistics does not provide satisfactorily, are techniques for first detecting patterns and regularities in data, without having to wait for an external impetus.

2   *A lack of confidence*

This manifests itself in different forms. At one extreme there is the free admission of a fear of numbers; at the other there are aggressive statements that management is solely a matter of instinct and that methodical evaluations of information are unnecessary ('if it can't be done on the back of an envelope . . . '). Whatever the manifestation, the effect is the same: little or no data analysis is done. In fact the majority of numbers problems require common sense and only a minimal knowledge of technical matters. Data analysis is rather like reading. When looking at a business report, a manager will usually read it carefully, work out what the author is trying to say and then decide whether he is correct. The process is similar with a table of numbers. The data have to be sifted, thought about and weighed. To do this, clear presentation (as proposed in the last chapter) is more important than sophisticated techniques. Most managers could do excellent data analyses provided they had the confidence to treat numbers as they would words. It is only because most people are less familiar with numbers than words that the analysis process needs to be made more explicit (as it will be later in the chapter) in the case of numbers.

3   *Over-complication by the experts*

The attitude of experts in numerical subjects can cause confusion. The experts use jargon which is fine when talking to their peers but not to laymen; they try sophisticated methods of analysis before simple ones; they communicate results in a complicated form, paying little regard to the users of the data, e.g. vast and indigestible tables of numbers as the output of a management information system. The result can be that the numbers experts within an organization distance themselves from day-to-day problems. Managers come to believe that they themselves have insufficient skills whilst at the same time they think that no realistic help is available from experts. Problems such as these are not, of course, confined to quantitative areas.

EXAMPLES OF THE PROBLEMS FACED

Accounting data

Table 2.1 shows the profit and loss account of a shipping company. It is difficult to analyse and understand what the significant features of the company's business were. In the last chapter this profit and loss account was re-presented, revealing that some important happenings in the company's financial year were obscured in Table 2.1.

Management information system (MIS) output

Table 2.2 is an example of perfectly accurate and meaningful data which are presented in a form convenient for the producer of the data but not the user. A manager would take a considerable time to understand the table. Since this is part of a MIS which contains many such tables, it is essential that a manager should be able to assimilate the information contained in the table very rapidly. A little thought given to simplifying the table in accordance with the needs of the user would bring an enormous improvement.

Market research

Table 2.3 indicates what can happen when experts over-complicate. The original data came from interviews of 700 television viewers who were asked which British television programmes they 'really like to watch'. The table is the result of the analysis of these relatively straightforward data. It is impossible to see what the real information is, even if one knows what correlation means. On the other hand a later and simpler analysis of the original data revealed a result of wide-ranging importance in the field of television research.

Guidelines for data analysis

There are five stages to the guidelines. They suggest a process for analysing a set or table of numbers. A manager who follows these steps should be better able to understand the data before him/her, or at least to see the deficiencies in the data and to know what more are needed.

Stage 1   Reduce the data

Many tables contain too many numbers. Data are often included on

Table 2.1   *Statement of revenue and expenditures*

		1981	1980
		£	1000 £
Gross Freight Income		98 898 684	73 884
Voyage Expenses:			
Own Vessels	31 559 336		(24 493)
Vessels on timecharter	42 142 838	(73 702 174)	(28 378)
Operating Expenses:			
Crew Wages and Social Security	7 685 965		(9 010)
Other Crew Expenses	541,014		(633)
Insurance Premiums	1 161 943		(1 367)
Provisions and Stores	1 693 916		(2 268)
Repairs and Maintenance	1 685 711		(3 297)
Other Operating Expenses	60 835		(27)
		(12 829 384)	
Result from Vessels		12 367 126	4 411
Results from Parts in Limited Partnerships		1 793 314	(163)
Result from Operation Business Building		167 343	0
Management Expenses			
Salaries, Fees, Employees' benefits etc	1 426 607		(1 208)
Other Management Expenses	502 815		(635)
		(1 929 422)	
Depreciation Fixed Assets		(7 106 305)	(5 365)
Transferred from Classification Survey Fund	112 401		652
Set aside for Classification Survey Fund	(612 401)		(27)
		(500 000)	
		4 792 056	(2 335)
Capital Income and Expenses:			
Misc. Interest Income	318 601		364
Misc. Interest Expenses	(8 450 307)		(6 482)
	(8 131 706)		
Net Profit/(Loss) currency exch	(190 836)		(680)
Dividends	35 732		47
		(8 286 810)	
Other Income and Expenses			
Net Profit from sales Fixed Assets	5 553 047		5 593
Balance Classification Survey and Self Insurance Fund upon Sale	1 414 132		858
	6 967 179		
Net Profit/(Loss) from sale of shares	(38 647)		0
Misc. Income/(Expenses) relating to previous years	629 630		1 408
Adjustm. cost shares	226 197		(206)
Loss from Receivables	(219 189)		0
Reversal write up Fixed Assets previous years	(2 026 067)		0
		5 539 103	
Result before Taxes and Allocations		2 044 349	(1 433)
Allocations		(1 929 400)	
		114 949	(1 433)
Reserves	(11 495)		1 071
Dispositions	(103 454)		362
		(114 949)	1 433

Table 2.2 Budgeting data from a MIS

PORT – BERGEN
PERIOD – DEC 1982

OCEAN PORT TERMINAL COSTS – SHIPSIDE OPERATIONS
(CURRENCY KRONER)

TERMINAL COSTS	MONTH				CUMULATIVE				
	ESTIMATE	STANDARD	VARIANCE	VAR %	ESTIMATE	STANDARD	VARIANCE	VAR %	BUDGET
LO-LO									
STEVEDORING									
STRAIGHT TIME – FULL	131223	143611	12388	8.6	1237132	1361266	124134	9.1	1564896
STRAIGHT TIME – M.T.	13387	14651	1264	8.6	256991	281399	24408	8.7	
(UN) LASHING	78		(78)		78		(78)		
SHIFTING	801		(801)		11594		(11594)		
OVERTIME, SHIFT TIME OF WAITING & DEAD TIME	7102		(7102)		190620		(190620)		
RO-RO									
STEVEDORING									
TRAILERS									
STRAIGHT FULL	20354	26136	5782	22.1	167159	215161	48002	22.3	330074
STRAIGHT M.T.	178	228	50	21.9	14846	18993	4147	21.8	
RO-RO COST PLUS VOLVO CARGO									
ROLLING VEHICLES	14326	19515	5189	26.6	98210	157163	58953	37.5	
BLOCKSTOWED	29	27	(2)	(7.4)	613	674	61	9.1	
(UN) LASHING RO-RO	355		(355)		355		(355)		
SHIFTING	977		(977)		3790		(3790)		
OVERTIME, SHIFT TIME OF WAITING & DEAD TIME	1417		(1417)		28713		(28713)		
HEAVY LIFTS (OFF STANDARD)					2009		(2009)		
CARS									
STEVEDORING									
STRAIGHT TIME	6127	6403	276	4.3	38530	35328	(3202)	(9.1)	168000
(UN) LASHING					2		(2)		
SHIFTING	795		(795)		1286		(1288)		
OVERTIME, SHIFT TIME DF WAITING & DEAD TIME					7573		(7573)		
OTHER SHIPSIDE OP COSTS	3422		(3422)		24473		(24473)		
TOTAL TERMINAL COSTS	200571	210571	10000	4.8	2083976	2069984	(13992)	(.7)	2062970

Table 2.3 Result of the analysis of some market research data. Adults who 'really like to watch': correlations to 4 decimal places (programmes ordered alphabetically within channel)

		PrB	ThW	Tod	WoS	GrS	LnU	MoD	Pan	RgS	24H
ITV	Prb	1.0000	0.1064	0.0653	0.5054	0.4741	0.0915	0.4732	0.1681	0.3091	0.1242
ITV	ThW	0.1064	1.0000	0.2701	0.1424	0.1321	0.1885	0.0815	0.3520	0.0637	0.3946
ITV	ToD	0.0653	0.2701	1.0000	0.0926	0.0704	0.1546	0.0392	0.2004	0.0512	0.2437
ITV	WoS	0.5054	0.1474	0.0926	1.0000	0.6217	0.0785	0.5806	0.1867	0.2963	0.1403
BBC	GrS	0.4741	0.1321	0.0704	0.6217	1.0000	0.0849	0.5932	0.1813	0.3412	0.1420
BBC	LnU	0.0915	0.1885	0.1546	0.0785	0.0849	1.0000	0.0487	0.1973	0.0969	0.2661
BBC	MoD	0.4732	0.0815	0.0392	0.5806	0.5932	0.0487	1.0000	0.1314	0.3267	0.1221
BBC	Pan	0.1681	0.3520	0.2004	0.1867	0.1813	0.1973	0.1314	1.0000	0.1469	0.5237
BBC	RgS	0.3091	0.0637	0.0512	0.2963	0.3412	0.0969	0.3261	0.1469	1.0000	0.1212
BBC	24H	0.1242	0.3946	0.2432	0.1403	0.1420	0.2661	0.1211	0.5237	0.1212	1.0000

the basis of reasoning such as 'well someone, somewhere may need them'. For the sake of these probably mythical people all others wanting to understand the data have to search for pearls in the dross. Each user has to distinguish what is important from a confusion of irrelevant data.

The producer of the data has to cater for the many requirements of the people who receive them. Even so, producers tend to err on the side of over-supply and include more than is necessary even for a wide range of users. In any case it is simpler to provide everything.

The first stage of an analysis is, then, to reduce the numbers. This means omitting those that are redundant or irrelevant. This is very much a matter of judgement, but it is important to resist the assumption that every single piece of data must be important because it has been included. The producer is unlikely to have included them because he knows they are important; he will probably have no idea whether they are important or not. He will have included them because they are available and he is being cautious. By reducing the data the analyst may well find that he is dealing with just a fraction of the original set.

Stage 2   Re-present the data

The visual approach is central to the understanding of numbers. A pattern can often be seen quickly in a set of well-arranged data which otherwise (for example when computerized) appears to be no more than a random jumble. Recall the similarity between handling words and numbers. In the same way that one will sift and meditate upon a verbal report, one needs to spend time thinking about a table of numbers. If the numbers are skilfully presented then the thinking process should be shorter and carry a greater chance of success. The visual aspect of data analysis and the need to present data well is usually neglected, perhaps because it does not have a high technical content, but the approach has been shown to work in practice.

The re-presentation being recommended does not refer just to data that are a random jumble. Data that have already been assembled in a neat table will still probably have to be re-presented (to oneself) in a fashion which makes it easy to see any patterns. The ways in which data can be rearranged were described in detail in the last chapter. There were 7 'rules' for data presentation:

(1) Round the numbers to 2 effective figures, e.g. 3521 becomes 3500, 352.1 becomes 350, 35.21 becomes 35, 3.521 becomes 3.5 and so on.

(2) Put rows and columns in size order.
(3) Interchange rows and columns so that the more important comparison is down rather than across.
(4) Use summary measures to focus rows and columns.
(5) Minimize use of space and grid lines.
(6) Make labelling clear without hindering the understanding of the numbers.
(7) Draw attention to important features by a verbal summary.

Stage 3   Build a model

Building a model is a mathematical sounding expression for a straightforward task. The objective is to find a pattern in the numbers which can be expressed simply and which can represent the whole set. It may just be some form of summary. The model may be a verbal one, an arithmetic one or an algebraic expression, either simple or complex. The following are examples of very different models, all of which proved adequate for the data and context concerned:

- row averages all equal but with ± 10% variation within each row;
- capital expenditure increased in real terms by 5% p.a. between 1970 and 1978 and decreased by 5% p.a. between 1979 and 1982;
- sales per salesman were approximately equal for each northern region; sales per salesman were approximately equal for each southern region – but the northern figure is 25% higher than the southern;
- column 1 (represented by y) is related to column 2 (represented by x) by the expression
  $$y = 2.71x + 3.34.$$

Frequently the simpler models prove sufficient. Even so, building a model can be difficult, requiring skill and creativity. It is at this stage that an analyst can be led along unnecessarily technical and mathematical routes.

The principal benefit of a model is that just a few numbers or words need to be handled instead of a profusion. If the model is a good one it can be used as a summary, for spotting exceptions, for making comparisons and for taking action.

The vast amount of information now available to managers (particularly in view of recent growth in the use of all sizes of computers) means that data analysis should be a speedy process. When looking for a model or pattern the simple ones should be investigated first. There is usually neither time nor sufficient specialist help to do otherwise. It has to be a simple model or nothing. More importantly, the simple models are usually the more useful since they are easier to handle and communicate. Only if the simple approach fails are complex methods necessary, and then expert knowledge may be required. However, it is worth noting that, even if the numbers appear to have no pattern, this is information of a sort and can be useful.

Stage 4 Exceptions

An exception can only exist if there is a general pattern from which it differs. Having found a pattern in some data, it is now meaningful to look at the exceptions to it. In managerial situations the exceptions can be more important than the pattern. For the marketing divisions of a company, the latest monthly sales figures may have the pattern that the sales volume per salesman is approximately the same in every sales region except for region x where the figure is lower. The exception can now be investigated. Did it occur because it is a large region geographically and thus more time is taken up travelling? Is it because someone is falling down on the job? Is it on account of reasons within or beyond the control of the company? A frequent reason for an extreme exception is that someone made a mistake with the numbers, for example a typist omitted a zero. This last possibility should always be considered.

Once the exceptions have been noticed they can be corrected, left alone (in cases of items beyond the company's control) or made the subject of management action. Whatever the process for investigating and dealing with exceptions, it is the recognition of a pattern or model in the numbers that makes it possible to see their existence.

It is a common error to deal with exceptions before discerning the pattern. This is the basis of the 'number picking' mentioned earlier. Not only is this illogical (since an exception must by definition have something to be an exception from), but it also leads to true exceptions being missed whilst spurious ones are discussed at length and in depth.

A further possibility may arise. The exceptions may be so numerous and/or inexplicable that the only conclusion has to be that the model chosen for the data is not good enough. It does not explain the fluctuations in the numbers sufficiently well to be regarded as a general model. In this situation the only option is to return to the data and attempt, in the light of knowledge of the inadequacies of the failed model, to find a new model which better represents the data. More iterations of this process may take place before a satisfactory model is found. As with verbal arguments, several re-drafts may be necessary.

Stage 5   Comparisons

Having detected the pattern and explained the exceptions to it, the results can be compared with other relevant information. Rarely is any statistical investigation undertaken completely in isolation. Nearly always, there are other results with which to make a comparison. The other results may be from another year, from another company, or from another country. In other words reference can usually be made to a wider set of information. In consequence, questions may be prompted: Why is the sales mix different this year from the previous five? Why do other companies have different levels of brand switching for their products? Why is productivity higher at similar French plants? Making comparisons such as these provides a context in which to evaluate the results of analysis, as well as suggesting the consistencies or anomalies which may require management action.

If the results coincide with others, then this further validates the model and may mean that in future fewer data may need to be collected — only enough to see whether the already established model still holds. This is especially true of Management Information Systems where managers receive regular print-outs of sets of numbers and are looking for changes in what has gone before. It is more efficient for the manager to carry an established model from one time period to the next rather than the raw data. It is easier to compare the models for the time periods than the original numbers.

EXAMPLE: CONSUMPTION OF DISTILLED SPIRITS IN THE USA

As an example of an analysis of numbers, Table 2.4 shows the consumption of distilled spirits in different states of the USA. The objective of the analysis would be to measure the variation in consumption across the states and to detect any areas where there were distinctive differences. How can the table be analysed and what information can be gleaned from it? Follow the five stages of the guidelines.

Table 2.4 Apparent consumption of distilled spirits (Comparison by states, calendar years 1976–75)

Estimated Population July 1, 1976	Percent to Total Population	Percent to U.S. Total Consumption 1976	1975	License States	Rank in Consumption 1976	1975	Consumption in Wine Gallons 1976	1975	Percent Increase Decrease	Per Capita 1976	Per Capita 1975
382,000	0.18	0.33	0.32	* Alaska	46	47	1,391,172	1,359,422	2.3	3.64	3.86
2,270,000	1.06	1.03	0.98	** Arizona	29	30	4,401,883	4,142,521	6.2	1.94	1.86
2,109,000	0.98	0.60	0.56	** Arkansas	38	38	2,534,826	2,366,429	7.1	1.20	1.12
21,520,000	10.03	12.33	12.32	** California	1	1	52,529,142	52,054,429	0.9	2.44	2.46
2,583,000	1.20	1.50	1.49	** Colorado	22	22	6,380,783	6,310,566	1.1	2.47	2.49
3,117,000	1.45	1.69	1.72	** Connecticut	18	18	7,194,684	7,271,320	(– 1.1)	2.31	2.35
582,000	0.27	0.35	0.36	** Delaware	45	43	1,491,652	1,531,688	(– 2.6)	2.56	2.65
702,000	0.33	1.08	1.14	** Dist. of Columbia	27	27	4,591,448	4,828,422	(– 4.9)	6.54a	6.74a
8,421,000	3.92	5.33	5.28	** Florida	2	4	22,709,209	22,329,555	1.7	2.70	2.67
4,970,000	2.31	2.52	2.35	* Georgia	13	13	10,717,681	9,944,846	7.8	2.16	2.02
887,000	0.41	0.48	0.47	* Hawaii	41	40	2,023,730	1,970,089	2.7	2.28	2.28
11,229,000	5.23	6.13	6.35	** Illinois	3	3	26,111,587	26,825,876	(– 2.7)	2.33	2.41
5,302,000	2.47	1.67	1.66	** Indiana	19	20	7,110,382	7,005,511	1.5	1.34	1.32
2,310,000	1.08	0.68	0.70	** Kansas	35	35	2,913,422	2,935,121	(– 0.7)	1.26	1.29
3,428,000	1.60	1.14	1.19	** Kentucky	26	26	4,857,094	5,006,481	(– 3.0)	1.42	1.47
3,841,000	1.79	1.66	1.59	** Louisiana	21	21	7,073,283	6,699,853	5.6	1.84	1.77
4,144,000	1.93	2.54	2.54	** Maryland	12	12	10,833,966	10,738,731	0.9	2.61	2.62
5,809,000	2.71	3.28	3.38	** Massachusetts	10	10	13,950,268	14,272,695	(– 2.3)	2.40	2.45
3,965,000	1.85	2.00	1.99	** Minnesota	15	15	8,528,284	8,425,562	1.2	2.15	2.15
4,778,000	2.23	1.66	1.82	* Missouri	20	17	7,074,614	7,679,871	(– 7.9)	1.48	1.61
1,553,000	0.72	0.64	0.64	** Nebraska	36	36	2,733,497	2,717,859	0.6	1.76	1.76
610,000	0.23	1.02	0.97	** Nevada	30	31	4,360,172	4,095,910	6.5	7.15	6.92
7,336,000	3.42	3.73	3.82	** New Jersey	8	8	15,901,587	16,154,975	(– 1.6)	2.17	2.21
1,168,000	0.54	0.47	0.46	** New Mexico	42	41	1,930,372	1,954,139	1.3	1.70	1.70
18,084,000	8.42	9.64	9.88	** New York	2	2	41,070,005	41,740,341	(– 1.6)	2.27	2.30
643,000	0.30	0.33	0.33	** North Dakota	47	46	1,388,475	1,384,311	0.3	2.16	2.18
2,766,000	1.29	0.92	0.99	** Oklahoma	33	29	3,904,574	4,187,527	(– 6.8)	1.41	1.54
927,000	0.43	0.49	0.50	** Rhode Island	39	39	2,073,075	2,131,329	(– 2.7)	2.24	2.30
2,848,000	1.33	1.39	1.26	** South Carolina	23	25	5,934,427	5,301,054	11.9	2.05	1.33
686,000	0.32	0.31	0.29	* South Dakota	48	48	1,312,160	1,242,021	5.6	1.91	1.82
4,214,000	1.95	1.32	1.27	** Tennessee	24	24	5,618,774	5,357,160	4.9	1.33	1.28
12,487,000	5.82	4.22	4.06	** Texas	5	6	17,990,532	17,167,560	4.8	1.44	1.40
4,609,000	2.15	2.56	2.54	*** Wisconsin	11	11	10,896,455	10,739,261	1.5	2.36	2.33
150,280,000	70.01	75.02	75.22	Total License	....	....	319,583,215	317,874,435	0.5	2.13	2.13

Stage 1   Reduce data

Many of the data are redundant. Are population figures really necessary when the table already contains consumption and per capita figures? Are percentage figures really necessary when per capita figures are given? The table can be reduced to a fraction of its original size with a minimal loss of real information.

Stage 2   Re-present

To understand the table more quickly, reduce the numbers to 2 effective figures. The original table has some numbers which have 8 figures. No analyst could possibly make use of this level of speci- fication. What conclusion would be affected if an eighth figure were, say, a 7 instead of a 4? In any event when one considers the way in which data such as these are probably collected, they surely cannot be accurate to 8 figures. If the table were a source document then more than 2 figures might be required, but not 8.

Putting the states in order of decreasing population is more help- ful than alphabetical order. Alphabetical order is useful for finding names in a long list, but it adds nothing to the analysis process. The new order means that states are just as easy to find (most people know that California has a large population and Alaska a small one, especially since no-one using the table will be totally ignorant of the demographic attributes of the states of the USA). At the same time, the new order makes it easy to spot states whose consumption is out of line with their population.

The end-result of these changes, together with some of a more cosmetic nature, is table 2.5. Contrast table 2.5 with the original, table 2.4.

Stage 3   Build a model

The pattern is evident when one looks at the amended table. Total consumption for each state varies with the population of the state. Per capita consumption in each state is about equal to the figure for all licence states (2.1 wine gallons) with just some variation (± 30%) about this level. The pattern a year earlier was the same except that overall consumption increases slightly (1%) between the 2 years. Refer back to table 2.4 and see if this pattern is evident even when you know that it is there.

Table 2.5   *Apparent consumption of distilled spirits*
(amended)

Licence states (in order of population)	Consumption		
	Wine gallons (million)		*Per capita 1976*
	1976	*1975*	
California	52.5	52.1	2.4
New York	41.1	41.7	2.3
Texas	18.0	17.2	1.4
Illinois	26.1	26.8	2.3
Florida	22.7	22.3	2.7
New Jersey	15.9	16.2	2.2
Massachusetts	14.0	14.3	2.4
Indiana	7.1	7.0	1.3
Georgia	10.7	9.9	2.2
Missouri	7.1	7.7	1.5
Wisconsin	10.9	10.7	2.4
Tennessee	5.6	5.4	1.3
Maryland	10.8	10.7	2.6
Minnesota	8.5	8.4	2.2
Louisiana	7.1	6.7	1.8
Kentucky	4.9	5.0	1.4
Connecticut	7.2	7.3	2.3
Sth Carolina	5.9	5.3	2.1
Oklahoma	3.9	4.2	1.4
Colorado	6.4	6.3	2.5
Kansas	2.9	2.9	1.3
Arizona	4.4	4.1	1.9
Arkansas	2.5	2.4	1.2
Nebraska	2.7	2.7	1.8
New Mexico	1.9	2.0	1.7
Rhode Island	2.1	2.1	2.2
Hawaii	2.0	2.0	2.3
D. Columbia	4.6	4.8	6.5
S. Dakota	1.3	1.2	1.9
N. Dakota	1.4	1.4	2.2
Nevada	4.4	4.1	7.2
Delaware	1.5	1.5	2.6
Alaska	1.4	1.4	3.6
All licence states	320.0	318.0	2.1

N.B. Total US population, 1 July 1976 = 214,659,000.

Stage 4   Exceptions

The overall pattern of approximately equal per capita consumption in each state allows the exceptions to be seen. From table 2.5, three states stand out as having a large deviation from the pattern. The states are District of Columbia, Nevada and Alaska. These states were exceptions to the pattern in the earlier year also. Explanations in the cases of District of Columbia and Nevada are readily found, being to do with the large non-resident populations. People live for a time in the state and consume distilled spirits there but are not included in the population figures (diplomats in DC, tourists in Nevada). Explanations for Alaska are not so evident, and a deeper investigation may be required. Whatever the explanations, the statistical method has done its job. The patterns and exceptions in the data have been found. Explanations are the responsibility of experts in the distilled spirits market of the USA.

Stage 5   Comparisons

A comparison between the two years is provided by the table. Other comparisons would help the analysis. The following data would be useful:

- earlier years, say, 5 and 10 years before;
- a breakdown of aggregate data into whisky, gin, vodka, etc.;
- other alcoholic beverages: wine, beer.

Having collected data from these other sources they would be analysed in the manner described, but the process would be shorter because the analyst would know what to look for. Care would be needed to ensure that like was compared with like; for example there should be an unchanging definition of consumption over the 10-year period.

Worked examples

1. *VITESSE ROYALE* ORGANIZATION OF MOTORISTS

This statement by a motoring organization was made in a national newspaper in 1981:

> Ten years ago the death rate on the roads of this country was 0.1 death for every 1 million miles driven. By 1975 a death

occurred every 12 million miles. Last year, according to figures just released, there were 6,400 deaths, whilst a total of 92,000 million miles were driven.

(a) Analyse these data.
(b) Was driving safer in 1980 than in 1970?
(c) What do you anticipate the death rate might be in 1985?
(d) What reservations do you have regarding your prediction?

(a) Analyse the data according to the guidelines.
    (1) Reduce the data by omitting some verbiage.
    (2) Re-present what is left, giving:

        1970: 1 death every 10 million miles
        1975: 1 death every 12 million miles
        1980: 6,400 deaths in 92,000 million miles
           = 1 death in every 14.4 million miles.

    (3) On average the ratio of deaths to miles driven has improved. The model of the data is that the average number of miles for each death increases by 20% every 5 years.
    (4) There are no exceptions to this model since the data are too few.
    (5) There is no other information with which to make comparisons but, if obtainable, road death data prior to 1970, casualities (as opposed to deaths), and other countries' road accident data, would be useful.

(b) It can be concluded that there was a reduction in the death rate on the roads. Whether driving is safer is a slightly different matter. The deaths include all road deaths and an improvement in the overall figure could mask a worsening for pedestrians. In any case safety is a personal thing, depending upon one's own circumstances. The overall rate for drivers may have improved because of improvements for long-distance drivers as a result of motorway construction. This would not make things safer for a predominantly town driver.

(c) If the model is correct, the number of miles for every death will increase by 20% over the next 5 years. In 1985 there would then be 1 death every 17.3 million miles.

(d) The reservations are that the model is based on:

    ● too few data;

- aggregate data which may obscure important trends in sub-sections of road users;
- purely time trends. This approach is superficial since it does not attempt to explain what underlying causes initiate the trends. Over 5 years these causes may change and disrupt the time trend.

2. NORTH ENGLAND WATER TRANSMISSION

Table 2.6 shows expenditure on water supply for different financial categories, in different regions. For instance, the Yorkshire region spent £106,154,000 in all, split into £1,270,000 on water resources, £51,462,000 on water supply, etc. Analyse the table, showing particularly, for each expenditure category, the regions that have unusual expenditure levels. Follow the five stages of the guidelines.

(1) Reduce the data by omitting the expenditures, but retaining the percentages. Conversely the percentages could be omitted and the expenditures retained. The percentages are preferred in view of the final objective of finding the unusual expenditure levels. The presence of both types of data merely confused the original table.

(2) Re-present the data by rounding the numbers to 2 effective digits. Re-order the regions by their total expenditures; re-order the expenditure categories by size. The result is the amended Table 2.7.

(3) The model for the data is that under each category of expenditure, the percentage for each region is roughly the same as the percentage for the total (the whole of North England). The model is to some extent presupposed by the question, since out-of-the-ordinary expenditures are sought.

(4) The major exceptions can be spotted by looking down each column and noting the regions whose percentage is very different from the percentage at the foot of the column.

(5) The only really valid comparison would have to be with previous expenditures.

Suppose a major exception is defined as being more than one-fifth away from the North England figure. For example, for 'Water supply' an exception is any region whose percentage for this category is not between $44 - 8.8$ and $44 + 8.8$ (8.8 being one-fifth of 44). The choice of one-fifth as the definition of an exception is a matter of

Table 2.6   *Water Finances: revenue expenditure (including supplementary depreciation) for year ending 31 December 1981*
(£'000)

	Total	Water resources	Water supply	Sewerage	Sewage treatment and disposal	Water quality regulation
North England	596,933 (100.0)	15,081 (2.5)	261,967 (43.9)	129,173 (21.6)	185,742 (31.1)	4970 (0.9)
Water authority areas:						
North West	144,463 (100.0)	1105 (0.8)	75,517 (52.2)	28,050 (19.4)	38,679 (26.8)	1112 (0.8)
Northumbrian	38,525 (100.0)	121 (0.3)	18,406 (47.8)	9963 (25.9)	9689 (25.1)	346 (0.9)
Severn–Trent	180,431 (100.0)	8401 (4.7)	68,861 (38.1)	39,136 (21.7)	62,301 (34.5)	1732 (1.0)
Yorkshire	106,154 (100.0)	1270 (1.2)	51,462 (48.5)	18,537 (17.5)	33,916 (31.9)	969 (0.9)
Anglian	127,360 (100.0)	4184 (3.3)	47,721 (37.5)	33,487 (26.3)	41,157 (32.3)	811 (0.6)

Table 2.7   Percentage revenue expenditure in water areas for year ending 31 December 1981 (percentages)

	Total	Water supply	Sewage treatment	Sewerage	Water resources	Water quality
Severn–Trent	100	38	34	22	4.7	1.0
North West	100	52	27	19	0.8	0.8
Anglian	100	37	32	26	3.3	0.6
Yorkshire	100	48	32	17	1.2	0.9
Northumbria	100	48	25	26	0.3	0.9
Total North England	100	44	31	22	2.5	0.8

Note: Because of rounding rows may not total exactly to 100%.

judgement. A fraction other than one-fifth would result in a tightening or a loosening of the definition.

The answer to the question posed is that the unusual exceptions are:

- water supply: no exceptions, but 3 regions are close to the limits;
- sewage treatment: no exceptions, but Northumbria is close to the limit;
- sewerage: Yorkshire – 2 other regions are nearly exceptions;
- water resources: there is a large variation in this category – no region is less than one-fifth away from the total percentage.
- water quality: Anglian, Severn-Trent.

Note that in this instance ordering the regions by size has had little effect on the analysis. This does not diminish the general usefulness of this rule. For instance, it might have been found that regions of a particular size were generally the exceptions.

Final comments

This chapter has presented guidelines intended to help managers who are not statistical specialists to analyse their data.

1   Reduce the data.
2   Re-present the data.
3   Build a model.
4   Investigate exceptions.
5   Compare.

If the problems that the guidelines are meant to counter are real, and if they do indeed prove useful in overcoming them, then there are implications for the producers of data, as well as the users. More attention should be paid to the uses to which the data are being put. This does not mean asking managers which data they want, since this will lead to an over-provision (what manager would say anything but that he needs vast quantities of data in order to take the important decisions that are his responsibility). Likewise, the practice of issuing all the data available results in a surfeit which merely confuses the aspiring analyst. Designers of information systems and writers of reports should consider exactly what contribution their data are supposed to make and how they are expected to improve decision-making. What is needed is a more systematic approach whereby data are provided not in isolation but in the context of the management processes they are intended to serve.

The second implication is more direct. Data should be presented in forms which enable them to be analysed speedily and accurately. Much of the reduction and re-presentation stages of the guidelines could, in most instances, be carried out just as well by the producer of the data as by the user. It would then need to be done only once rather than many times by the many users of the data. Unfortunately, if time is spent thinking about the presentation of statistics, it is usually spent in making the tables look neat or attractive rather than making them amenable to analysis.

As for the managers themselves, few will not admit that there are currently problems with the provision and analysis of data, but they rarely tell this to their computer manager or whoever sends them data. Without feedback, inexpensive yet effective changes are never made. It is the responsibility of users to criticize constructively the form and content of the data they receive. The idea that computer scientists/statisticians always know best, and that if they bother to provide data then they must be useful, is false. The users must make clear their requirements, and even resort to a little persistence if alterations are not forthcoming.

Of course it is difficult to make general prescriptions about the handling of numbers because every manager sees the problem differently. He sees it mainly in the (probably) narrow context of his own work. One manager sees numbers only in the financial area, another sees them only in production management. Even so, the guidelines suggested here are intended to be generally applicable to the analysis of management data in many different situations and with a range of different requirements. The key points are:

- in most situations managers without statistical backgrounds can carry out satisfactory analyses themselves;
- the use of simple methods is preferable to complex ones;
- visual inspection of well-arranged data can play a role in coming to understand them;
- data analysis is like verbal analysis — the guidelines described here merely make explicit for numbers what is done naturally when dealing with words.

The need for better skills to turn data into real information in managerial situations is not new. What has made the need so urgent in recent times is the exceedingly rapid development of computers and associated management information systems. The ability to provide vast amounts of data very quickly has grown enormously. It has far outstripped the ability of management to make use of the data. The result has been that in many organizations managers have been swamped with so-called information which in fact is no more than just numbers. The problem of data analysis is no longer a small one that can be ignored. When companies are spending large amounts of money on data provision, the question of how to turn the data into information and use them in decision-making is one that has to be faced.

Further examples

1. GENERAL AUTOMOTIVE SUPPLIERS

Table 2.8 is part of the information system of a petroleum company. The purpose of the system is to provide managers with general background information. The table shows the imports of coal by five countries during the years 1975–79. Analyse the data, stating what the main pattern and exceptions are.

Table 2.8   Hard coal imports (tonnes)

	1975	*1976*	*1977*	*1978*	*1979*
Belgium	523,478	608,765	489,088	593,223	752,167
Denmark	338,611	341,255	462,921	506,433	449,769
Italy	1,103,516	936,730	971,386	977,144	868,886
Netherlands	350,711	400,467	397,410	399,601	484,477
Spain	315,945	373,228	369,103	411,405	431,154

2. SOCIAL COUNCIL FOR RESEARCH INTO EMPLOYMENT AND WAGES

Table 2.9 was taken from the report on a research project conducted under the sponsorship of the Council. The project was intended to find out the extent to which employers were paying wages which were lower than nationally agreed minimum wages. It involved a sample survey of 11 trades in 4 geographical areas. For each of the trades the table shows; (a) percentage of those employers sampled who were found to be underpaying; (b) percentage of employees

Table 2.9 Wages underpayment

Trades	Proportion of employers inspected who were found to be underpaying	Proportion of employees' wages examined that were entitled to arrears	Amount underpaid (£)
Retail Bread	31.99	21.22	12,516.01
Bookselling	50.00	20.07	11,205.48
Drapery and Outfitting	32.85	19.20	41,922.26
Retail Food	22.74	10.89	26,850.33
Furnishing and Allied Trades	21.19	9.86	14,085.62
Newsagency and Tobacco	33.93	20.63	19,564.66
Hairdressing	26.23	10.30	4,394.04
Licensed Non-Residential Hotels	23.01	11.24	6,218.02
Licensed Restaurants	30.77	11.79	13,939.20
Unlicensed Restaurants	48.58	34.78	15,315.83
Other Trades	18.86	10.20	10,157.30
Total	28.08	14.91	176,168.75

included in the survey who were being underpaid; (c) the total amount underpaid in the year of the survey in each trade by the employers included in the sample. Analyse the data and suggest what the main conclusions might be. If no general pattern exists, suggest what extra data are required to further the analysis.

Answers

1. General Automotive Suppliers

The main pattern is that 4 of the countries' imports have risen by between 30% to 45% over the years 1975–79, although the increases have not been even from year to year. The fifth country, and the exception, is Italy where coal imports have decreased by about 20% over the years in question.

2. Social Council for Research into Employment and Wages

No major pattern emerges, but there is some indication that trades for which a large percentage of employers are underpaying also has a large percentage of employees being underpaid. There are, however, many exceptions to this model. The column referring to the amount underpaid is meaningless unless one knows the sample size, i.e. the number of employers and employees concerned in the amounts underpaid. This fact suggests that the next step in the analysis might be to ascertain the sample sizes and thereby calculate the average amounts underpaid per employee in each trade. This is a case where, even though no neat analysis emerges, it is clear what to do next.

CHAPTER 3

Summary measures

By the end of the chapter the reader should know when summaries are useful, how to calculate them and how to choose between different measures. Outliers in a data set can distort the calculations and should be dealt with. Indices are a special case of summary measures.

When trying to understand and remember the important parts of a lengthy verbal report it is usual to summarize. This may be done by expressing the essence of the report in perhaps a few sentences, or by underlining key phases, or by listing the main subsection headings. Each individual has his own method which may be physical (a written précis) or mental (some way of registering the main facts in the mind). Whatever the method, the point is that it is easier to handle information in this way, by summarizing and storing only brief summaries in the memory. On the few occasions that details are required it is necessary to return to the report itself.

The situation is no different when numerical rather than verbal data are being handled. It is still better to form a summary. The summary may be a pattern, simple or complex, revealed when analysing the data, or it may be based on one or more of the standard summary measures described in this chapter.

Inevitably some accuracy is lost. In the extreme, if the summarizing is badly done it can be wholly misleading (how often do people claim to have been totally misunderstood after hearing someone else's summary of their work?). Take the case of a recent labour strike in the UK. The strike was about wages payment. In stating the current levels of payment newspapers/union leaders/employers could not, of course, report the payments for all the 120,000 employees in the industry. They had to summarize. Figure 3.1 shows five statements as to the 'average' weekly wage.

'The average weekly wage is . . . '

Quote 1: '£57' (union leader)
Quote 2: '£78' (newspaper A)
Quote 3: '£92' (newspaper B)
Quote 4: '£115' (newspaper C)
Quote 5: '£138' (employers' organization)

Figure 3.1

All these quoted wages were said to be the same thing, the 'average weekly wage'. Are the employees in the industry grossly underpaid or overpaid? It is not difficult to choose an amount that reinforces one's prejudices. The discrepancies are not because of miscalculations but because of definitions. Quote 1 is the basic weekly wage without overtime, shift allowances and unsocial hours allowance, and it has been reduced for tax and other deductions. Since the industry is one which requires substantial night-time working for all employees, no-one actually takes home the amount quoted. Quote 2 is the same as the first but without the tax deduction. Quote 3 is the average take-home pay of a sample of 30 employees in a particular area. Quote 4 is of basic pay plus unsocial hours allowance but without any over-time or tax deductions. Quote 5 is basic pay plus allowances plus maximum overtime pay, without tax and other deductions.

It is important when using summary measures (and in all of statistics) to apply common sense and not be intimidated by complex calculations. Just because something which sounds statistical is quoted ('the average is £41.83') it does not mean that its accuracy and validity should be accepted without question. When summary measures fail it is usually not because of poor arithmetic or poor statistical knowledge but because common sense has been lacking.

In this context the remainder of the chapter goes on to describe ways of summarizing numbers, and discusses their effectiveness and their limitations.

Table 3.1   Ship unloading data: monthly report

	THIS MONTH: JUNE			PREVIOUS MONTH: MAY			
	DATE	TONNAGE	WEEKLY TOTAL		DATE	TONNAGE	WEEKLY TOTAL
MON	1	1240					
TUES	2	740					
WED	3	900					
THURS	4	1140					
FRI	5	1125	5145	FRI	1	1238	1238
MON	8	1038		MON	4	0	
TUES	9	650		TUES	5	1250	
WED	10	842		WED	6	1051	
THURS	11	941		THURS	7	956	
FRI	12	728	4199	FRI	8	842	4099
MON	15	1226		MON	11	1045	
TUES	16	831		TUES	12	540	
WED	17	933		WED	13	938	
THURS	18	720		THURS	14	847	
FRI	19	915	4625	FRI	15	739	4109
MON	22	1130		MON	18	1049	
TUES	23	934		TUES	19	844	
WED	24	825		WED	20	741	
THURS	25	825		THURS	21	847	
FRI	26	820	4534	FRI	22	736	4217
MON	29	1134		MON	25	0	
TUES	30	830	1964	TUES	26	1050	
				WED	27	946	
				THURS	28	741	
				FRI	29	735	3472
JUNE TOTAL			20,467	MAY TOTAL			17,135

The usefulness of summary measures

There are several *types* of summary measure. Each type of measure summarizes a different aspect of the data. For management purposes it is usually possible to summarize adequately a set of data using at most three types of measure. For example, the managers of a general cargo port facility receive monthly a computer report of the daily tonnage unloaded. The report for June (22 working days) is as in table 3.1. The data as shown are useful for reference purposes (for example, what was the tonnage on the 15th?) or for the background to a detailed analysis (for example, is tonnage always higher on a Monday and, if so, by how much?). Both these types of use revolve around the need for detail. For more general information purposes (for example, was May a good month? what is the trend this year?) the amount of data contained in the table is too large and unwieldy for the manager to be able to make the necessary comparisons. It would be rather difficult to gauge the trend of tonnage levels so far this year, from six reports, one for each month, such as that in table 3.1. If summary measures were provided then most questions, apart from the ones that require detail, could be answered readily. A summary of table 3.1 might be as shown in table 3.2.

Table 3.2   Summary report of June/May tonnage

	June	May
Average tonnage/day	930	900
Range of daily tonnage	650–1240	540–1250
Shape of distribution	Symmetrical	Symmetrical

Data exclude holidays when there was zero tonnage.

The summary information provided in table 3.2 enables a wide variety of management questions to be answered and, more importantly, answered quickly. Comparisons are made much more easily if summary data for several months, or years, are available on one report. The full details of table 3.1 would still be available for those who needed them.

Three types of summary measure are used in table 3.2. The first, average production, measures the 'location' of the numbers and tells at what general level the data are. The second, the range of production, measures 'scatter' and indicates how widely spread the data are. The third, the symmetry of the data, measures the 'shape' of the data. In this case the answer 'symmetrical' says that the data fall equally on either side of the average.

The three measures reflect the important attributes of the data. No important general features of the data are omitted. If, on the other hand, the measure of scatter had been omitted, 2 months could have appeared similar whereas by virtue of having very different ranges (say, month 1: 950–1050, month 2: 500–1300) an important piece of information reflecting production planning problems would be lost.

For each *type* of measure (location, scatter, etc.) there is a choice of measure to use (for location, the choice is between the arithmetic mean and other measures). The measures have applications other than as summaries. They also form essential parts of various theories. For example the variance, a measure of scatter, plays a central role in modern financial theory.

Measures of location

Measures of location are intended to show, in general terms, the size of a set of data.

ARITHMETIC MEAN

The most common and useful measure is the well-known arithmetic mean. It is defined:

$$\text{Arithmetic} = \frac{\text{Sum of readings}}{\text{Number of readings}}$$

This can be put in shorthand (or mathematical) notation:

$$\bar{x} = \frac{\Sigma x}{n}$$

where: x refers to the data in the set;
$\bar{x}$ (x bar) is standard notation for the arithmetic mean;
Σ is the Greek capital sigma and, mathematically, means 'sum of';
n is standard notation for the number of readings in the set.

For example, the nine numbers:

$$3, 4, 5, 5, 5, 6, 7, 8, 11$$

$$\bar{x} = \frac{3 + 4 + 5 + 5 + 5 + 6 + 7 + 8 + 11}{9}$$

$$= \frac{54}{9}$$

Arithmetic mean = 6

Note that the term 'average' usually refers to the mean, unless otherwise stated.

MEDIAN

This is the middle value of a set of numbers. There is no mathematical formula for calculating it. It is obtained by listing the numbers in ascending order and then the median is that number at the halfway point.

For example, using the same nine numbers as above which are already in ascending order:

$$3, 4, 5, 5, 5, 6, 7, 8, 11$$

↑
Middle
number

Median = 5

Note that the median is not the middle point of the range. It is not halfway between 3 and 11, but is the number that has as many readings above as below.

If there is an even number of readings then there can be no one middle number. In this case it is usual to take the arithmetic mean of the middle two numbers as the median. For example, if the set of nine numbers of above was increased to ten by the presence of

'12' the set would become:

$$3, 4, 5, 5, 5, 6, 7, 8, 11, 12$$

$$\uparrow\uparrow$$

Middle
two
Numbers

$$\text{Median} = \frac{5 + 6}{2} = 5.5$$

MODE

The third measure of location is the mode. This is the most frequently occurring value. Again, there is no mathematical formula for the mode. The frequency with which each value occurs is noted and the value with the highest frequency is the mode. Again, using the same nine numbers as an example:

$$3, 4, 5, 5, 5, 6, 7, 8, 11$$

Number	Frequency
3	1
4	1
5	3
6	1
7	1
8	1
11	1

Mode = 5

FURTHER EXAMPLES OF MEASURES OF LOCATION

When a set of numbers is plotted with the values on the horizontal axis and the frequencies of those values on the vertical axis, the resulting graph is called a *histogram*.

Figures 3.2(a), (b) and (c) show three different sets of data together with their histograms. Each has a different shape of histogram. Check your understanding of definitions by calculating mean, median and mode for each set. Which measure of location is the best for each set? Table 3.3 shows the answers.

Context: Marks out of fifteen scored by each of twenty participants in a driving competitition. A symmetrical distribution.
Readings: 5, 5, 6, 6, 7, 7, 7, 7, 8, 8, 8, 8, 8, 9, 9, 9, 10, 10, 11, 12.
Shape: See figure 3.2(a).

(a)

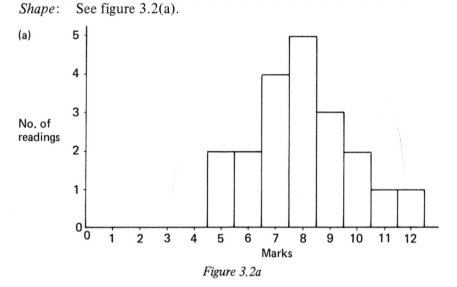

Figure 3.2a

Context: Number of episodes in a 19 part serial seen by a sample of 20 viewers. A 'U'-shaped distribution.
Readings: 0, 0, 0, 0, 0, 1, 1, 1, 1, 2, 2, 4, 17, 18, 18, 19, 19, 19, 19, 19.
Shape: See figure 3.2(b).

(b)

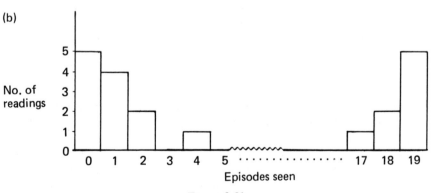

Figure 3.2b

Context: Weeks away from work through sickness in a 1-year period for a sample of 20 employees in a particular company. A reverse J-shaped distribution.
Readings: 0, 0, 0, 0, 0, 0, 0, 1, 1, 1, 1, 2, 2, 2, 3, 5, 18, 28, 44, 52.
Shape: See figure 3.2(c).

(c)

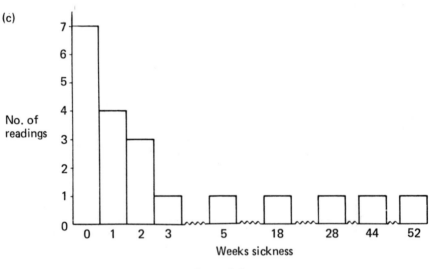

Figure 3.2c

Table 3.3   *Answers for calculations of Figure 3.2*

Readings	Mean	Median	Mode
Symmetrical	8.0	8	8
U-Shape	8.0	2	0 and 19
Reverse-J	8.0	1	0

Note that the arithmetic mean of all three distributions is the same, 8.0, despite the data being very different, indicating the need for care in summarizing data.

For the symmetrical distribution, mean, median and mode are equal. This is always approximately the case for symmetrical data. Whichever measure is chosen, a similar answer results. Consequently it is best to use the most well-known measure, the arithmetic mean, to summarize location for this set of data.

The U-shaped data are typical for certain TV series and also magazine readership where many people see hardly any programmes/ issues, many people see all or most, and just a very few fall between the two extremes. Here the arithmetic mean is not very helpful, since no-one actually sees 8 episodes. As a summary of the data, it would mislead. The median is not a particularly good summarizer either. Data in which virtually all readings are at one of two extremes cannot easily be reduced to a single measure which is the middle value of the set. The median is also very sensitive to small changes in this case. A serial that was only slightly more popular could have resulted in a median of around 18. Best of all is the mode. In this case there are two, 0 and 19. Note that even when technically there is only one mode for the data, it is usual, when there is more than one definite peak, to quote more than one mode, each corresponding to one peak. Reporting that there are two modes will capture the essential nature of the data, its U shape.

The reverse-J shaped distribution arises with distributions that are truncated at one end (no values less than 0) but which have a few outliers at the other end of the scale. It is typical of sickness records where most workers are absent very little or not at all, but where one or two employees with a major illness may be absent for most of the year. The best measure is the median. It tells how many weeks off the 'middle' employee had. There are major defects for the other two measures. The arithmetic mean is distorted by the outliers. A very different value would be obtained if one of the outliers were not present. For example, if the 52 were omitted the arithmetic mean would be $(108/19) = 5.7$. In all situations the arithmetic mean can be misleading if there are one or two extremes in the data. The mode is not misleading, just unhelpful. Most sickness records have a mode of 0, therefore to quote 'mode 0' is not providing any more information than merely saying that the data concern sickness records. Non-symmetrical data are said to be skewed. These sickness data are skewed to the right because the extreme values are to the right of the rest of the distribution.

Mean, median and mode are the major measures of location and summarizers of data, but they do not capture all aspects of a set of numbers. Other types of summary measure are necessary. However, before leaving measures of location, their uses, other than as summarizers, will be described.

OTHER USES OF MEASURES OF LOCATION

(a) *As a focus for the eye*

8,7,5,11,10,7,8,8,6,7,10,12,5,6,7,9,9,8,8,9
2,1,0,18,5,0,52,2,1,0,1,44,3,0,0,28,0,0,2,1

Figure 3.3

The importance of the visual approach in data analysis has already been stressed. One aspect of this is the use of summary measures of location as a focus to guide the eye and give a more speedy impression of the meaning of data. Consider the two sets of data in figure 3.3. It is difficult to see the shape or pattern in them. In figure 3.4 the mean of each set of data has been added and, as the eye goes along the row, the shape of the distribution becomes apparent quickly. In the first case the focus enables one to see that the numbers are scattered closely and about equally either side of the mean; in the second case one sees that most numbers are below the mean with just a few considerably above.

8,7,5,11,10,7,8,8,6,7,10,12,5,6,7,9,9,8,8,9   Mean 8.0
2,1,0,18,5,0,52,2,1,0,1,44,3,0,0,28,0,0,2,1   Mean 8.0

Figure 3.4

In fact the two sets are the symmetrical data and the reverse-J-shaped data introduced earlier. In the latter case the arithmetic mean was judged not to be the most useful measure to act as a summary; nevertheless it has a value when used as a focus for the eye. One meets this usage with row and column averages in tables of numbers.

(b) *For comparisons*

Measures of location can be used to compare two (or more) sets of data regardless of whether the measure is the best summary measure for the set. Figure 3.5 shows two sets of data containing a different

Set 1   5,7,8,9,9,10   Mean = 8
Set 2   5,5,5,6,6,7,8,10   Mean = 6.5

Figure 3.5

number of readings. The arithmetic mean may or may not be the correct summary measure for either set. Nevertheless, a useful comparison between them can be affected through the mean. Similarly, the sickness records of a group of people (reverse-J shape) over several years can be compared using the arithmetic mean, even though one would not use this measure purely to summarize the data.

THE IMPORTANCE OF THE ARITHMETIC MEAN

For each *type* of summary measure, several actual measures are available for use. In the case of measures of location, the possibilities are arithmetic mean, median and mode. Which measure should be used in which situation? What are the strengths and weaknesses of each measure? The choice between arithmetic mean, median and mode is simpler than it might be. The reason is that the arithmetic mean is pre-eminent. It is easy to calculate and to use and it is widely understood and recognized. The arithmetic mean, therefore, is always used . . . unless there are good reasons not to. Here are three examples of good reasons.

Distortion of the mean by outliers

The arithmetic mean is sensitive to large outliers. This was the case with the reverse-J-shaped distribution. Whenever large outliers are present, use of the *median* should be considered.

Another example of this effect would be in calculating the average salary in a small engineering firm where one finds:

		Salary(inc. bonuses)
1	Founder/Managing Director	£40,000
4	Skilled workmen	£ 8,000
5	Unskilled workmen	£ 6,000
	Arithmetic mean salary =	£10,200

The average salary as calculated is totally unrepresentative of salaries earned in the firm, purely because of the distorting presence of the high earnings of the Managing Director. The median salary is £7000 and thus more representative.

Distortion of the mean by clusters

The arithmetic mean can be unrepresentative if the data split into two or more distinctly different clusters. The mean may well fall in the middle and be some distance from any piece of the recorded data. Such is the case for the U-shaped distribution described earlier referring to TV viewing. People see almost none or almost all the episodes. The mean was calculated as coming between the two clusters at the extremes. In such cases the *mode* may be more valuable.

Taking averages of averages

When an average of an average is taken, the arithmetic mean can be incorrect. Consider the following case of 2 streams of 15-year-old pupils at a high school and their average marks in a national examination.

Stream A: 50 pupils, average mark 74%
Stream B: 30 pupils, average mark 50%

What is the average mark for both streams together? It is tempting to say

$$\text{Overall average } \frac{74 + 50}{2} = 62\%.$$

However this figure is the average of two averages. To get the true average, one has to go back to first principles:

$$\text{Overall average} = \frac{\text{Sum of marks for all pupils}}{\text{Number of pupils}}$$

$$= \frac{\overset{\text{Stream A}}{(50 \times 74)} + \overset{\text{Stream B}}{(30 \times 50)}}{50 + 30}$$

$$= \frac{3700 + 1500}{80}$$

$$= \frac{5200}{80}$$

$$= 65\%$$

The lesson is: when averaging averages where groups of different size are involved go back to the basic definition of the average.

Except where one of the difficulties described above applies, the arithmetic mean is the first-choice measure of location.

Measures of scatter

Measures of scatter do exactly as their name implies. They are a measure of the extent to which the readings are grouped closely together or are scattered over a wide interval. They are also called measures of dispersion.

RANGE

The best-known and certainly simplest measure of scatter is the range which is the total interval covered by the numbers.

$$\text{Range} = \text{Largest reading} - \text{Smallest reading}$$

For example, the nine numbers

$$3,4,5,5,5,6,7,8,11$$
$$\underline{\text{Range} = 11 - 3 = 8}$$

INTERQUARTILE RANGE

The range is defined entirely on the two extreme values and this might be misleading. To overcome this problem the interquartile range is the range of the numbers after eliminating the highest and lowest 25%. This measure is no longer sensitive to extremes.

$$\text{Interquartile range} = \text{range of the middle 50\% of readings}$$

For example,

$$3,4,5,5,5,6,7,8,11$$

The top 25% of readings (= approximately the top 2 numbers) and the bottom 25% (= approximately the bottom 2 numbers) are eliminated.

3,4,5,5,5,6,7,8,11
↑↑   ↑↗
remove   remove

$$\underline{\text{Interquartile range} = 7 - 5 = 2}$$

MEAN ABSOLUTE DEVIATION

A measure which involves all the readings is the mean absolute deviation (MAD). It is the average distance of the readings from their arithmetic mean and as such is an intuitively reasonable way of quantifying scatter.

$$MAD = \frac{\text{sum (difference between each reading and mean)}}{\text{number of readings}}$$

Or, more mathematically

$$MAD = \frac{\Sigma |x - \bar{x}|}{n}$$

where $\bar{x}$ is the arithmetic mean
     n is the number of readings in the set
    | | is notation for 'ignoring the sign'.

The notation | | (said 'the absolute value of') means that whether the reading is above or below the mean, its distance from the mean is measured as a positive number. For example, calculate the MAD of

$$3,4,5,5,5,6,7,8,11$$

From previous work, $\bar{x} = 6$

x	3, 4, 5, 5, 5,6,7,8, 11
$x - \bar{x}$	−3,−2,−1,−1,−1, 0, 1, 2, 5
$\|x - \bar{x}\|$	3, 2, 1, 1, 1, 0, 1, 2, 5
$\Sigma\|x - \bar{x}\| =$	3 + 2 + 1 + 1 + 1+0+1+2+5
	= 16

$$MAD = \frac{16}{9} = 1.8\dot{}$$

The concept of absolute value used in the MAD is to overcome the fact that $(x - \bar{x})$ is sometimes positive, sometimes negative, and sometimes zero. The absolute value gets rid of the sign. Why is it necessary to get rid of the sign? Try the calculations without taking absolute values and see what happens.

VARIANCE

An alternative way of eliminating the sign from deviations from the mean is to square them since the square of any number is never negative. The variance is the average squared distance of readings from the arithmetic mean.

$$\text{Variance} = \frac{\text{Sum of squares of deviations of each reading from the mean}}{\text{Number of readings} - 1}$$

Or, more mathematically,

$$\text{Variance} = \frac{\Sigma(x - \bar{x})^2}{n - 1}$$

There is a technical reason for the divisor being $n - 1$, rather than n. Refer to the bibliography for an explanation.

Taking the usual example, calculate the variance of

$$3, \ 4, \ 5, \ 5, \ 5, \ 6, 7, 8, 11$$

x	3,	4,	5,	5, 5, 6, 7, 8, 11				
$x - \bar{x}$	$-3,$	$-2,$	$-1,$	$-1,-1, 0, 1, 2, 5$				
$(x - \bar{x})^2$	9,	4,	1,	1, 1, 0, 1, 4, 25				

$$\Sigma(x - \bar{x})^2 = 9 + 4 + 1 + 1 + 1 + 0 + 1 + 4 + 25$$

$$= 46$$

$$\text{Variance} = \frac{\Sigma(x - \bar{x})^2}{n - 1}$$

$$= \frac{46}{8} = 5.75$$

The variance has many theoretical uses, particularly in financial theory. However, as a pure description of scatter it suffers from the disadvantage that it is in squared units. The variance of the number of weeks sickness of 20 employees would have units of square weeks.

STANDARD DEVIATION

The standard deviation attempts to overcome this problem by taking the square root of the variance. It is therefore in the same units as the original data. It also has important theoretical applications of its own.

$$\text{Standard deviation} = \sqrt{\text{Variance}}$$

Or, in mathematical form,

$$\text{Standard deviation} = \sqrt{\frac{\Sigma(x - \bar{x})^2}{n - 1}}$$

In the example, variance = 5.75.

Therefore, standard deviation = $\sqrt{5.75}$ = 2.4

COMPARING THE MEASURES OF SCATTER

In the case of measures of location, the choice between measures is relatively easy. The arithmetic mean is pre-eminent and it is chosen unless there is a good reason to do otherwise. For measures of scatter, the relative merits of the measures are more evenly balanced and the choice is more difficult. Table 3.4 shows the major strengths and weaknesses of each measure.

Table 3.4   Comparison of measures of scatter

Measure	Advantages	Disadvantages
Range	Easily understood; familiar	Based on just two observations; grossly distorted by outliers; descriptive measure only
Interquartile range	Easily understood	Not well known; descriptive measure only
Mean absolute deviation	Intuitively sensible	Unfamiliar; difficult to handle mathematically
Variance	Mathematically tractable	Wrong units; no intuitive appeal
Standard deviation	Mathematically tractable; well-known because of its use in various theories	Too involved for descriptive purposes

All the measures have their particular uses. No one stands out. When a measure of scatter is required purely for descriptive purposes the best measure is probably the mean absolute deviation, although it is not as well-known as it deserves to be. When a measure of scatter is needed as part of some wider statistical theory, then the variance and standard deviation are frequently encountered.

Example

A company's 12 salesmen in a particular region last month claimed the following amounts in expenses.

Salesman	Expenses (£)
1	370
2	490
3	300
4	330
5	390
6	390
7	360
8	460
9	350
10	480
11	430
12	330

Calculate: (a) range;
             (b) interquartile range;
             (c) MAD;
             (d) variance;
             (e) standard deviation.

Which measure is the best for summarizing the scatter in these data?

(a) Range = highest − lowest = 490 − 300 = £190

(b) Putting the numbers in ascending order
     300,330,330, | 350,360, 370,390,390,430, | 460,480,490
             ↑   ↑
          lowest   highest
          quartile   quartile

     Interquartile range = 430 − 350 = £80

(c) To calculate MAD, it is first necessary to find the arithmetic mean:

$$\text{Mean} = \bar{x} = \frac{4680}{12} = 390$$

Next calculate the deviations:

x	370	490	300	330	390	390	360	460	350	480	430	330
$x - \bar{x}$	−20	+100	−90	−60	0	0	−30	+70	−40	+90	+40	−60
$\lvert x - \bar{x}\rvert$	+20	+100	+90	+60	0	0	+30	+70	+40	+90	+40	+60

$$\text{MAD} = \frac{\Sigma\lvert x - \bar{x}\rvert}{n} = \frac{600}{12} = £50$$

(d) The variance

$x - \bar{x}$	−20	+100	−90	−60	0	0	−30	+70	−40	+90	+40	−60
$(x - \bar{x})^2$	400	10,000	8100	3600	0	0	900	4900	1600	8100	1600	3600

$\Sigma(x - \bar{x})^2 = 42,800$

$$\text{Variance} = \frac{\Sigma(x - \bar{x})^2}{n - 1}$$

$$= \frac{42,800}{11}$$

$$= \underline{3890}$$

(e) Standard deviation $= \sqrt{\text{Variance}}$

$$= \sqrt{3890}$$

$$= \underline{£62.4}$$

The best descriptive measure of scatter in this situation is the mean absolute deviation. The average difference between a salesman's expenses and the average expenses is £50. This is a sensible measure which involves all data points. The range is of great interest, but not as a measure of scatter. Its interest lies in indicating the discrepancy between the highest and lowest expenses. It says nothing about the ten in-between salesmen. The interquartile range is probably the second choice. The variance and standard deviation are too complex conceptually to be descriptive measures in this situation where further statistical analysis is not likely.

Dealing with outliers

The data from which summary measures are being calculated may well include one or more unusual observations which may have a disproportionate effect on the result. How does one deal with the outliers? Are they to be included or excluded? Three basic situations arise.

Twyman's law

This only half-serious law states that any piece of data that looks interesting or unusual is wrong. The first consideration when confronted by an outlier is that the number is incorrect, perhaps because of an error in collection or a typing mistake. There is an outlier in these data which are the week's overtime payments to a group of 7 workmen.

13.36, 17.20, 16.78, 15.98, 1432, 19.12, 15.37

Twyman's Law suggests that the outlier showing a payment of 1432 occurs because of a dropped decimal point rather than a fraudulent claim. The error should be corrected and the number retained in the set.

Part of the pattern

An outlier may be a definite and regular part of the pattern and should neither be changed nor excluded. Such was the case with the sickness record data of figure 3.2 (c). The outliers were part of the pattern and similar effects were likely to be seen in other time-periods and other groups of employees.

Isolated events

Outliers occur which are not errors but which are unlikely to be repeated; i.e. they are not part of the pattern. Usually they are excluded from calculations of summary measures, but their exclusion is noted. For example, the following data, recorded by trip-wire, show the number of vehicles travelling down a major London road during a 10-day period:

5271, 5960, 6322, 6011, 7132, 5907, 6829, 741, 7098, 6733

The outlier is the 741. Further checking shows that this day was the occasion of a major royal event and that the road in question was closed to all except state coaches and police vehicles. This is an isolated event, perhaps not to be repeated for several years. For traffic control purposes the number should be excluded from calculations since it is misleading, but a note should be made of the exclusion. Hence one would report:

$$\text{Mean vehicles/day} = \frac{5271 + 5960 + 6322 + 6011 + 7132 + 5907 + 6829 + 7098 + 6733}{9}$$

$$= \frac{57{,}263}{9}$$

$$= 6{,}363*$$

* Excluding day of royal event.

The procedure for outliers is first to look for mistakes and correct them; second, to decide whether the outlier is part of the pattern and should be included in calculations or an isolated event and should be excluded.

Indices

An index is a particular form of measure used for summarizing the movement of a variable over time. An indexed series should be easier to understand and compare with other series. The best-known type of index is probably a cost of living index. The cost of living comprises the cost of many different goods: foods, fuel, transport, etc. Instead of the miscellaneous and confusing prices involved for all these purchases an index number summarizes it all. If the index for 1982 is 182.1 compared with 165.3 in 1981, it can be calculated that the cost of living has risen by

$$\frac{182.1 - 165.3}{165.3} \times 100\%$$

$$= 10.2\%$$

This is rather easier than having to cope with the range of individual price rises involved.

Every index has a *base year* when the index was 100. If the base year for the cost of living index was 1976, then the cost of living has risen 82.1% between 1976 and 1982. It could be said in a different way: the 1982 cost of living is 182.1% of its 1976 value.

The index very quickly gives a feeling of what has happened to the cost of living. Comparisons also are easier. For example, if the wages and salaries index had 1976 as its base year and stood at 193.4 in 1982, it can be seen that over the 6 years wages outstripped the cost of living, 93.4% as against 82.1%.

The definition of a cost of living index can be quite complicated. At the other end of the scale there are some more basic indices.

SIMPLE INDEX

At the most primitive level an index is just the result of the conversion of one series of numbers into another based on 100. Consider the average price of a new dwelling in each of 10 years. Choosing the base year arbitrarily as 1978, the original series and the index are as in Table 3.5. The index for 1978 is 100, being the base year. The other data in the series are scaled up or down accordingly. For instance, the index for 1975 is:

$$8.6 \times \frac{100}{12.4} = 69$$

↑   ↖

Original   Original datum
datum   for base year
for 1975

And for 1981:

$$19.7 \times \frac{100}{12.4} = 159$$

The choice of the base year is important. It should be such that individual index numbers during the time span being studied are never too far away from 100. As a rule of thumb, the index numbers are not usually allowed to differ from 100 by more than a factor of 3. The numbers should be in the range 30 → 300. If the base year for the numbers in Table 3.5 had been chosen as 1973, then the index series would have been from 100 to 318.

In long series there might be more than one base year. For example, a series covering more than 30 years from 1950 to 1982 might have 1950 as a base year with the series then rising to 291 in 1969, which could then be taken as a second base year (see table 3.6). Care obviously needs to be taken in interpreting this series. The

Table 3.5

	1973	1974	1975	1976	1977	1978	1979	1980	1981	1982
Price (£ thousands)	6.1	8.2	8.6	10.1	11.8	12.4	16.9	19.0	19.7	19.4
Index	49	66	69	81	95	100	136	153	159	156

Table 3.6

Year	1950	1969	1982
Index	100	291	
		100	213

increase from 1950 to 1982 is not 113%. If the original series had been allowed to continue the 1982 index would have been 213 x 2.91 = 620. The increase is thus 520%.

SIMPLE AGGREGATE INDEX

The usefulness of an index is emphasized when it is summarizing several functions. A monthly index for meat prices could not be based on monthly prices for, say, beef alone. It would have to take into account prices of other types of meat. An aggregate index does just this. Table 3.7 shows the prices per kilo at market for beef, pork and lamb. A simple aggregate combines the different prices by adding them up and forming an index from the total. For instance, the total price for January is

$$148 + 76 + 156 = 380$$

Table 3.7   *Meat prices (p per kilo)*

	Cattle	*Pigs*	*Lamb*
Jan	148	76	156
Feb	150	80	167
Mar	155	75	180
Apr	163	79	194
May	171	81	186
June	179	76	178
July	184	82	171
Aug	176	76	168
Sept	142	79	163
Oct	146	84	160
Nov	149	87	159
Dec	154	94	162

For February it is

$$150 + 80 + 167 = 397$$

If January is taken as the base, then the indices for the months are as shown in table 3.8. One possible disadvantage of this index is that livestock with a low price will have much less influence than livestock with a high price. For instance, in February a 20% change in the price of cattle would change the index much more than a 20% change in the pig price.

February: $150 + 80 + 167 = 397$ Index = 104.5

With 20% cattle price change $180 + 80 + 167 = 427$ Index = 112.4

With 20% pig price change $150 + 96 + 167 = 413$ Index = 108.7

This may be a desirable feature. If the price level of each factor in the index reflects its importance then the higher-priced elements should have more effect on the index. On the other hand this feature may not be desirable. One way to counter this is to construct a *price-relative index*. This means that the individual prices are first converted into an index and then these individual indices averaged to give the overall index.

Table 3.8

Month	Total	Index
Jan	380	100
Feb	397	104
Mar	410	108
Apr	436	115
May	438	115
June	433	114
July	437	115
Aug	420	111
Sept	384	101
Oct	390	103
Nov	395	104
Dec	410	108

WEIGHTED AGGREGATE INDEX

The question of whether to use prices or price relatives in the index described above is really all about weighting. How much influence should the individual parts have, relative to one another, on the overall index? This issue is of greatest importance in a cost of living index. A simple aggregate index would not be a suitable method of combining the prices of milk, bread, fruit, tobacco, electricity and so on. A weighted aggregate index allows different weights to be given to the different prices. What the weights should be is still a matter of analysis and judgement, but in the case of price indices the quantities purchased are often used.

Returning to the livestock index, suppose the quantities purchased at the livestock markets from which the price data were obtained, are shown in table 3.9 (quantities in thousand tonnes). The prices are first weighted by the quantities. The final index is formed from the

Table 3.9

	Cattle		Pigs		Lamb	
	Price	*Quantity*	*Price*	*Quantity*	*Price*	*Quantity*
Jan	148	1.6	76	0.25	156	0.12
Feb	150	1.5	80	0.23	167	0.11
Mar	155	1.4	75	0.25	180	0.10

resulting monthly total. The quantities used for the weighting should be the same for each month, since this is a price index. Otherwise the index would measure price *and* volume changes. If the quantities used for weighting are the base month (January) quantities, then the index is known as a *Laspeyres Index* and is calculated as shown in table 3.10. If the base month (or year) were a long time ago, or if the quantities in the base month were in some way unusual, then

Table 3.10

Month	Weighted total	Index
Jan	$(148 \times 1.6) + (76 \times 0.25) + (156 \times 0.12) = 274.5$	100
Feb	$(150 \times 1.6) + (80 \times 0.25) + (167 \times 0.12) = 280.0 \quad \dfrac{280.0}{274.5} \times 100 = 102$	
Mar	$(155 \times 1.6) + (75 \times 0.25) + (180 \times 0.12) = 288.3 \quad \dfrac{288.3}{274.5} \times 100 = 105$	

the weights can be taken from the most recent time-period. The resulting index is known as a *Paasche index*. In the livestock example, a Paasche index would weight the prices in each month with the quantities relating to December, the most recent month. A Paasche index always uses the most up-to-date weightings, but it has the serious disadvantage in practice that every time new data arrive (and the weightings change) the entire series of past indices also changes.

Most frequently a *fixed-weight index* is used, where the weights are from neither the base period nor the most recent. They may be from some intermediate period or from the average of several periods. It is a matter of judgement to decide which fixed weights to use.

For a cost of living index the situation is even more difficult. People's purchasing behaviour changes over time in response to price levels, wages, technology, etc. For instance, more alcohol, more petrol, more meat, but less fish, less bread is purchased now than 20 years ago. These changes should be reflected in the index, which should be based on the range of things an average family purchases. In other words when comparing the cost of living index now with that of 20 years ago the indices should have different weights as well as the inevitably different prices. In practice, with nationally published indices of this type, weights are changed at regular intervals in recognition of changes in purchasing behaviour, but the changes do not affect past values of the index. The timing and level of the changes is usually a matter of discussion and judgement based on research data.

Worked examples

1. HOME PRODUCTS

Home Products are a mail-order firm selling household goods through catalogues. They operate from a large combined office block and warehouse. Their expenditure on light bulbs is very high and they would like to reduce it. An experiment was recently carried out to test the light bulbs of two different suppliers. Fifty bulbs from each supplier were tested to destruction. The results shown in table 3.11 were obtained for the length of life.

Table 3.11   Length of life (hours)

	800–899	900–999	1000–1099	1100–1199	1200–1299	Total
Supplier I	10	12	14	8	6	50
Supplier II	5	22	19	3	1	50

(a) Which supplier's bulbs have greater average length of life?
(b) Which supplier's bulbs are more uniform in quality?
(c) How would the answers to (a) and (b) be used in deciding which supplier's bulbs to use?

In calculating summary measures from categorized data, approximations have to be made. Each category is replaced by its mid-point. For example, all bulbs failing after 800—899 hours are regarded as having failed at 850 hours.

(a) Mean life for Supplier I bulbs

$$= \frac{(10 \times 850) + (12 \times 950) + (14 \times 1050) + (8 \times 1150) + (6 \times 1250)}{50}$$

$$= \frac{51,300}{50}$$

$$= \underline{1026 \text{ hours}}$$

Mean life for Supplier II bulbs:

$$= \frac{(5 \times 850) + (22 \times 950) + (19 \times 1050) + (3 \times 1150) + (1 \times 1250)}{50}$$

$$= \frac{49,800}{50}$$

$$= \underline{996 \text{ hours}}$$

(b) To measure uniformity of quality means measuring the scatter of length of life. Range and interquartile range do not depend on all 50 observations. Since two suppliers are being compared and the differences may be small, a more precise measurement is required. Standard deviation (or variance) could be used, but mean absolute deviation is easier to calculate.

For Supplier I:

$$\text{MAD} = \frac{\begin{array}{c} 10 \times |850 - 1026| + 12 \times |950 - 1026| + 14 \times |1050 - 1026| \\ + 8 \times |1150 - 1026| + 6 \times |1250 - 1026| \end{array}}{50}$$

$$= \frac{5344}{50}$$

$$= \underline{107 \text{ hours}}$$

For Supplier II:

$$\text{MAD} = \frac{\begin{array}{c}5 \times |850 - 996| + 22 \times |950 - 996| + 19 \times |1050 - 996| \\ + 3 \times |1150 - 996| + 1 \times |1250 - 996|\end{array}}{50}$$

$$= \frac{3484}{50}$$

$$= \underline{70 \text{ hours}}$$

(c) Having a longer average life is obviously a desirable characteristic. However, uniform quality is also desirable.

As an example, consider a planned maintenance scheme in which bulbs (in factories or offices) are all replaced at regular intervals whether they have failed or not. This can be cheaper since labour costs may dominate bulb costs and replacement all at once takes less labour time than individual replacement. In such a scheme the interval between replacement is usually set such that, at most, perhaps 10% of bulbs are likely to have failed. The interval between replacement can be larger for bulbs of more uniform quality since they all tend to fail at about the same time. Consequently, the scatter of length of life can be more important than average length of life in these circumstances.

The choice between Supplier I and Supplier II depends upon the way in which the bulbs are replaced and is not necessarily based solely on average length of life.

2. S. HARP

Peter Harris is a salesman for the Harp building materials company. He travels regularly to all parts of the country. At the end of the last quarter he was called to the Accountant's office to explain his expenses record. In the quarter, Harris had made 10 trips (see table 3.12). The Accountant told Harris that his expenses were too high, because the average expense per day was £40 (£400/10). Other salesmen away on weekly trips submitted expenses which averaged £30/day. How should Harris argue his case?

The Accountant is averaging averages. The last column is the average expense per day for each trip. This is being averaged to give an 'average expense per day' of £40 (= £400/10). It would be more

Table 3.12

Trip	Days	Expenses (£)	Expenses per day (£)
1	3	102	34
2	1	48	48
3	8	160	20
4	7	147	21
5	½	35	70
6	6	144	24
7	2	90	45
8	½	30	60
9	1	50	50
10	3	84	28
TOTALS	32	890	400

reasonable to calculate the true average expense per day (as opposed to the average daily expense per trip):

$$\text{Average expense per day} = \frac{\text{Total expense}}{\text{Total days}}$$

$$= \frac{£890}{32}$$

$$= £27.8$$

i.e. less than that of the other salesmen.

An even better measure would be one that related expense to *both* trips and days, since short trips (of which Harris has several) are more expensive on a per-day basis.

3. SOUTHERN OIL

Table 3.13 shows the prices and sales totals for 3 petroleum products of Southern Oil. Prices are given in pence per gallon and production figures in millions of barrels.

(a) Use 1977 = 100 to construct a simple aggregate index for the years 1977–79 for the prices of the 3 petroleum products.

Table 3.13

	Prices			Quantities		
	1977	*1978*	*1979*	*1977*	*1978*	*1979*
Product I	26.2	27.1	27.4	746	768	811
Product II	24.8	28.9	26.5	92	90	101
Product III	23.0	24.1	24.8	314	325	348

(b) Use 1977 quantities as weights and 1977 = 100 to construct a weighted aggregate index for the years 1977 to 1979 for the prices of the 3 petroleum products.

(c) Does it matter which index is used? If so, which one should be used?

(d) How else could the index be constructed?

Index for 1977 = 100 for both types of index.

(a) Simple aggregate index

$$\textit{For 1978} \quad \text{Index} = \frac{27.1 + 28.9 + 24.1}{26.2 + 24.8 + 23.0} \times 100$$

$$= \frac{80.1}{74.0} \times 100$$

$$= \underline{108.2}$$

$$\textit{For 1979} \quad \text{Index} = \frac{27.4 + 26.5 + 24.8}{26.2 + 24.8 + 23.0} \times 100$$

$$= \frac{78.7}{74}$$

$$= \underline{106.4}$$

(b) Weighted aggregate index

$$\textit{For 1978} \quad \text{Index} = \frac{(27.1 \times 746) + (28.9 \times 92) + (24.1 \times 314)}{(26.2 \times 746) + (24.8 \times 92) + (23.0 \times 314)} \times 100$$

$$= \frac{30,442.8}{29,048.8} \times 100$$

$$= \underline{104.8}$$

For 1979

$$\text{Index} = \frac{(27.4 \times 746) + (26.5 \times 92) + (24.8 \times 314)}{(26.2 \times 746) + (24.8 \times 92) + (23.0 \times 314)} \times 100$$

$$= \frac{30,665.6}{29,048.8}$$

$$= \underline{105.6}$$

(c) Summary:

	1977	1978	1979
Simple index	100	108.2	106.4
Weighted index	100	104.8	105.6

It does matter which is used since between 1978 and 1979 the indices move in different directions. The weighted index must be the better one since it allows for the fact that very different quantities of each product are purchased.

The importance of the weighting is two-fold. Firstly, petroleum products are all made from crude oil and are to some extent substitutable as far as the producer is concerned. An index should reflect the average price of every gallon of petroleum product purchased. Only the weighted index does this. Secondly, products such as Product I, of which large quantities are purchased, will have a bigger effect on the general public than products for which small amounts are purchased. Again, the index should reflect this.

(d) A differently constructed index would use different weightings. The price part of the calculation could be changed by using price relatives but this would have little effect since the prices are close together. The different weightings that could be used are:

(1) Most recent year quantity weighting. But this would imply a change in historical index values every year.
(2) Average quantity for all years weighting. This would not necessarily mean historical changes every year. The average quantities for 1977–79 could be used for several years. This also has the advantage that it guards against the chosen base year being untypical in any way.

Final comments

In the process of analysing data, at some stage the analyst tries to form a model of the data. Pattern or summary are close synonyms for model. The model may be simple (all rows are approximately equal) or complex (the data are related via a multiple-regression model). Often specifying the model requires intuition and imagination. At the very least, summary measures can provide a model based on specifying for the data set:

- number of readings;
- a measure of location;
- a measure of scatter;
- the shape of the distribution.

In the absence of other inspiration, these four attributes provide a useful model of a set of numbers. If the data consist of two or more distinct sets (as, for example, a table) then this basic model can be applied to each. This will give a means of comparison between the rows or columns of the table or between one time-period and another.

The first attribute (number of readings) is easily supplied. Measures of location and scatter have already been discussed. The shape of the distribution can be found by drawing a histogram and literally describing its shape (as with the symmetrical, U, and reverse-J distributions seen earlier). A short verbal statement about the shape is often an important factor in summarizing or forming a model of a set of data.

Once a summary of whatever form exists, the manager has the means of comparing, checking, noticing trends and noticing exceptions. Summary data would be a useful addition to management information system print-outs. They would suggest to the manager the detail he need look at and the detail he could avoid. They would also enable him to compare quickly information relating to different time periods.

Further examples

1. EASTERN REGION RAILWAY

The railway company has collected data showing the extent to which trains have been on time in the last 3 months:

Minutes Late	0–1	2–4	5–9	10–14	15–19	20–29	30–39	40+
Percentage of trains	41	21	12	8	5	5	3	5

Summarize these data.

2. PAN RUSSIAN AIR NAVIGATION GROUP

A small charter aircraft belonging to the Group has to make a round trip, as shown in figure 3.6. The trip starts at A. Stops are made at B, C and D before returning to A. Because of prevailing weather conditions, the speed on each leg of the journey is estimated as shown. What is the average speed for the whole trip?

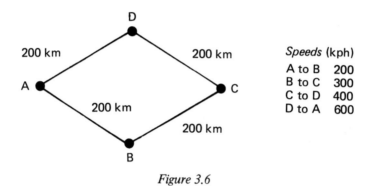

Speeds	(kph)
A to B	200
B to C	300
C to D	400
D to A	600

Figure 3.6

3. FINANCIAL ANALYSIS DATA SERVICES

The risk associated with a portfolio of investments can be gauged from the volatility of the percentage return it gives. Table 3.14 shows the return measured at the end of each week of a 13-week period for 3 portfolios. Calculate the variance for each, and thereby suggest which portfolio would seem to carry the most risk.

Table 3.14   Return percentage at each week-end

	1	2	3	4	5	6	7	8	9	10	11	12	13
Portfolio I	11.7	10.6	9.4	10.1	11.0	11.7	13.1	13.4	12.9	12.5	12.2	11.8	11.7
Portfolio II	9.7	9.1	8.7	9.5	10.9	11.9	12.8	12.1	11.3	10.2	10.9	10.2	9.2
Portfolio III	9.1	9.3	9.4	9.2	9.1	9.0	9.3	9.7	9.6	9.8	9.6	9.5	9.6

ANSWERS

1. Eastern Region Railway

Median time late = 2 minutes approximately. Quarter of the trains are more than 10 minutes late.

2. Pan Russian Air Navigation Group

Average speed (arithmetic mean) = 320 kph.

3. Financial Analysis Data Services

	Mean	Variance
Portfolio I	11.7	1.4
Portfolio II	10.5	1.6
Portfolio III	9.4	0.1

PART II

Statistical Methods

Part I was concerned with handling sets of numbers so that the information contained in them might be understood. Part II introduces some traditional statistical methods. These are formal methods for extracting particular pieces of information from a set of numbers. They are less generally applicable than the ideas of Part I, but they are more rigorous and precise. Although the methods may not be practised frequently by managers, the concepts on which they are based allow numerical information to be viewed more constructively and rationally.

The reader's prime objective should be to understand the concepts.

Preparation

Chapters 4 and 5 are at a higher technical level. There are sections which are left as 'black boxes'. Certain facts will have to be accepted without proof in order to avoid high-level mathematics. Before starting, the non-numerate reader must read Appendix A. This explains the idea of probability which is fundamental to both chapters. Throughout the chapters there are sections headed 'Technical note' that the non-numerate reader might wish to miss out the first time through.

Statistical distributions

By the end of the chapter the reader should be aware of how and why distributions, especially standard distributions, can be useful. Having numbers in the form of a distribution helps both in describing and analysing. Distributions can be formed from collected data; or they can be derived mathematically from knowledge of the situation in which the data are generated. The latter are called standard distributions. Three standard distributions – the normal, the binomial and the Poisson – are discussed in the chapter.

Several examples of distributions have been encountered already. In the last chapter, driving competition scores formed a symmetrical distribution; viewings of TV serial episodes formed a U-shaped distribution; sickness records formed a reverse-J-shaped distribution. These distributions, particularly their shapes, were used as part of the *description* of the data. A closer look will now be taken at distributions and their use in the *analysis* of data.

There are two general types of distribution: *observed* and *standard* (also called *theoretical* or *probability*). An observed distribution derives from collecting data in some situation. The distribution is peculiar to that situation. A standard distribution is derived mathematically. It should, theoretically, be applicable to all situations exhibiting the appropriate mathematical characteristics. Standard distributions have the advantage that time and effort spent on data collection is saved. The formation and application of both types will be described in this chapter.

Observed distributions

An observed distribution starts with the collection of numbers. The numbers are all measurements of a *variable*. A variable is just what the word implies. It is some entity which can be measured and for which the measurement varies when observations are made of it. For instance, the variable might be the consumption of distilled spirits in different states of the USA or the gross domestic product of Japan over several years. Figure 4.1 shows a collection of numbers which are observations of some variable. It is a mess. At first most data are. They may have been taken from dog-eared production dockets or mildewed sales invoices. They may be the output of a computer system where no effort has been made at data communication. Some sorting out must be done. A first attempt might be to arrange the numbers in order as in Figure 4.2. This is an *ordered array*. The numbers look neater now but it is still difficult to get a feel for the data – the average, the variability, etc. – as they stand. The next step is to classify the data. Classifying means grouping the numbers in bands to make them easier to appreciate. Each class has a frequency which is the number of data points that fall within that class. A *frequency table* is shown in figure 4.3. This shows that 7 data points (or observations or readings) were greater than or equal to 40 but

53

41   66
                              106
                    71
123   99
          83   92
    72
                    99
                                          22
              20
    78

Figure 4.1

	52	59	66
.	54	60	66
41	55	60	.
43	56	60	.
45	57	61	.
46	57	62	.
48	58	62	.
49	58	63	
49	58	65	
50	59	65	

Figure 4.2

Class	Frequency
$40 \leqslant x < 50$	7
$50 \leqslant x < 60$	12
$60 \leqslant x < 70$	22
$70 \leqslant x < 80$	27
$80 \leqslant x < 90$	19
$90 \leqslant x < 100$	10
$100 \leqslant x < 110$	3

NB $\leqslant$ means 'less than or equal to'
$<$ means 'less than'

Figure 4.3

less than 50; 12 were greater than or equal to 50 but less than 60; and so on. It is now easier to get an overall conception of what the data mean. Most of the numbers are between 60 and 90 with extremes of 40 and 110. Another arrangement, a *frequency histogram*, has even greater visual impact (figure 4.4).

Figure 4.4 makes it easy to see that the data are spread symmetrically over a range from 40 to just over 100 with the majority falling around the centre of the range. As a *descriptive* device the frequency histogram is satisfactory without further refinement. If, on the other hand, there were *analytical* objectives then it may be developed into a *probability histogram*. This is done from the connection between frequencies and probabilities. The probability that any randomly selected measurement lies within a particular class interval can be calculated:

P (number lies in class x) = Frequency class x/Total frequency

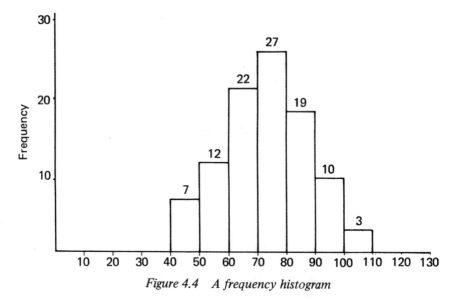

Figure 4.4   A frequency histogram

The frequency histogram becomes a probability histogram by making the units of the vertical axis probabilities instead of frequencies. The shape of the histogram remains the same. Once the histogram is in probability form, it is referred to as a *distribution*, although to be strictly accurate all configurations in figures 4.1 to 4.4 are distributions. Analyses, based on the manipulation of probabilities, can now be carried out. For example, what is the probability that any number will be in the range 50 to 70? From figure 4.4:

$$P(50 \leqslant x < 60) = 0.12$$
$$P(60 \leqslant x < 70) = 0.22$$

These probabilities are calculated from frequencies of 12 and 22 respectively for each class. The frequency for the combined class 50 to 70 is therefore 34 (= 12 + 22). Consequently:

$$\underline{P\,(50 \leqslant x < 70) = 0.34}$$

As a further example, what is the probability that any number is less than 70?

$$P\,(x < 70) = P\,(x < 50) + P\,(50 \leqslant x < 60) + P\,(60 \leqslant x < 70)$$
$$= 0.07 + 0.12 + 0.22$$
$$= \underline{0.41}$$

The calculations can be made in either frequency or probability form. Probability calculations such as these form the basis of analyses with observed distributions.

Example

Table 4.1 shows the number of deliveries of food made to a hyper-market (number of trucks arriving at the hypermarket) per day over a 1-year period (300 working days). The frequencies have been turned into percentages and then probabilities. For instance on 53 days there were between 0 and 9 deliveries and therefore the probability of having this number of deliveries on any day is $53/300 = 0.18$ (approx.). Figure 4.5 shows a probability histogram of the same data.

(a) The facilities at the hypermarket are only sufficient to handle 39 deliveries per day. Otherwise overtime working is necessary. What is the probability that on any day overtime will be required?

From Figure 4.5:

$$P \text{ (deliveries 70 or more)} = 0.02$$
$$P \text{ (deliveries 60–69)} \quad = 0.03$$
$$P \text{ (deliveries 50–59)} \quad = 0.04$$
$$P \text{ (deliveries 40–49)} \quad = 0.07$$

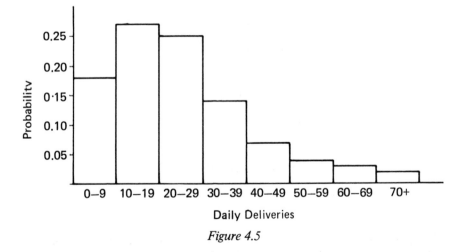

Figure 4.5

Table 4.1   Number of deliveries per day

Quantity demanded	No. of days	Percentage	Probability
0– 9	53	18	0.18
10–19	82	27	0.27
20–29	75	25	0.25
30–39	42	14	0.14
40–49	21	7	0.07
50–59	12	4	0.04
60–69	9	3	0.03
70 +	6	2	0.02
	300	100	1.00

Therefore,

$$P \text{ (deliveries exceed 39)} = 0.02 + 0.03 + 0.04 + 0.07$$

$$= 0.16$$

There is a 16% chance that overtime will be needed on any day.

(b) The capacity can be increased from the present 39 by taking on more staff and buying more handling equipment. To what level should the capacity be raised if there is to be no more than a 5% chance of overtime being needed on any day?

A capacity level x is to be found such that:

$$P \text{ (deliveries exceed } x) = 0.05$$

Since:

$$P \text{ (deliveries 70 or more)} = 0.02$$

$$P \text{ (deliveries 60–69)} \quad = 0.03$$

Then:

$$P \text{ (deliveries exceed 59)} = 0.02 + 0.03$$

$$= 0.05$$

Therefore, $x = 59$

A capacity of 59 will result in overtime being needed on 5% of days.

Standard distributions

Observed distributions often entail a great deal of data collection. Not only must sufficient data be collected for probabilities to be calculated and the distribution to take shape, but also the data must be collected individually for each and every situation. *Standard* distributions can reduce the amount of data collection.

A standard distribution is one that has been defined mathematically from a theoretical situation. The characteristics of the situation were expressed mathematically and the resulting distribution theoretically constructed by calculating the probability of the variable taking different values. When an actual situation, resembling the theoretical, arises, the associated standard distribution is applied. For example, one standard distribution, the *normal*, was derived from the theoretical situation of a variable being generated by a process which should give the variable a constant value, but does not do so because it is subject to many small disturbances. As a result the variable is distributed around the central value. The mathematical implications of such a situation mean that the probabilities of different values and the shape of the distribution could be anticipated. The normal distribution would then apply, for instance, to the lengths of machine-cut steel rods. The rods should all be of the same length, but are not because of variations introduced by vibration, machine setting, the operator, etc. The normal distribution applies to many situations with characteristics such as the above. It will be discussed in detail below. There are other standard distributions which derive from situations with other characteristics.

In summary therefore, using an *observed distribution* implies that data have been collected and histograms formed; using a *standard distribution* implies that the situation in which the data are being generated resembles closely a theoretical situation for which a distribution has been constructed mathematically.

Normal distribution

The most-used and best-known standard distribution is the normal.

CHARACTERISTICS

The normal distribution is bell-shaped and symmetrical (figure 4.6). It is also *continuous*. The distributions met up to now have been

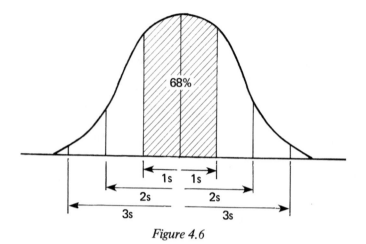

Figure 4.6

discrete. A distribution is discrete if the variable takes distinct values such as 1,2,3 . . . but not those in between. The delivery distribution (table 4.1) is discrete since the data are in groups. Continuous means that all values are possible for the variable. It could take the values 41.576, 41.577, 41.578 . . . rather than all these values being grouped together in the class 40–49 or the variable being limited to whole numbers only. An implication of continuity is that probabilities are measured not by the height of the distribution above the x-axis (as in the discrete case), but by areas under the curve. For example $P(65 < x < 75)$ is the area under the part of the curve between $x = 65$ and $x = 75$. The reason for this is mathematical, but intuitively it can be seen that since $P(x = 63.629732)$, say, must be infinitesimally small, a continuous distribution in which probabilities were measured by the heights of the curve above points (as with the discrete distributions) would be virtually flat.

The normal distribution also has important attributes with respect to the standard deviation (see figure 4.6):

68% of area (i.e. 68% of readings) lie within ± 1 standard deviation of mean

95% of area (i.e. 95% of readings) lie within ± 2 standard deviations of mean

99.7% of area (i.e. 99.7% of readings) lie within ± 3 standard deviations of mean

It is known that the IQ (intelligence quotient) of children is normally distributed with mean 100 and standard deviation 17. The attributes of figure 4.6 imply that:

68% children have an IQ in range 83–117
(mean ± 1 standard deviation)

95% children have an IQ in range 66–134

99.7% children have an IQ in range 49–151

Tables exist which give the area under any part of a normal curve, not just within ranges specified by whole numbers of standard deviations. Table 4.3, which will be explained later, is an example.

SITUATIONS IN WHICH THE NORMAL OCCURS

The normal distribution is constructed mathematically from the following theoretical situation. Repeated observations are taken of the same quantity L. Each time an observation is taken the quantity is subject to many sources of variation. Each source gives rise to a change in the value of L by $+u$ or $-u$ (u is some very small amount). It is a matter of chance and equally likely whether the variation is positive or negative. The variations are independent of one another (no variation is influenced by any other) and can be added together.

Suppose there were 100 such sources. If, on the first observation 57 variations say, acted positively and 43 negatively, the quantity measured would be $L + 57u - 43u = L + 14u$. The second observation might have 38 acting positively, 62 negatively. The quantity measured would be $L - 24u$. Over many observations a distribution of values would be obtained. Because of positives and negatives cancelling each other the tendency would be for most values to be close to L, but with a few at greater distances. Intuitively, the resulting symmetrical bell-shape can be visualized.

Mathematically the formula for the distribution is constructed from this theoretical situation assuming that the sources of variation are large in number and small in magnitude. From the mathematical formula the areas under all parts of the distribution are calculated and tables formed (table 4.3).

A great many practical situations approximate very well to the theoretical. It is not necessary to know what the sources of variation are, merely that they are likely to exist. Actual situations in which the normal distribution has been found to apply include:

- IQs of children.
- Heights of people of the same sex.

- Dimensions of mechanically produced components.
- Weights of machine-produced items.
- Arithmetic means of large samples.

In the case of human beings the variations giving rise to normal distributions of IQs, heights and other variables are presumably many genetic and environmental effects. For mechanically produced items the sources must include vibration (from a number of factors), tiny differences in machine setting and different operators. The use of the distribution in sampling is one of the most important uses. The variations arise from the random choice of what each sample comprises. This will be covered in the next chapter.

PARAMETERS

The normal distribution applies to both IQs and weights of bread loaves, yet the shapes of the two distributions are different (see figure 4.7). The distributions differ because the parameters differ.

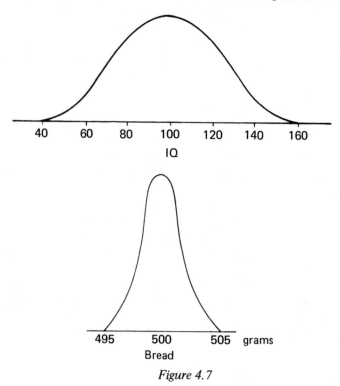

Figure 4.7

The normal distribution has two parameters, the arithmetic mean and the standard deviation. Two normal distributions with the same mean and the same standard deviation will be identical; two normal distributions with different means and standard deviations, while still having the characteristics shown in figure 4.6, will be centred differently and be of different widths. Parameters fix the context within which the variable varies. In terms of the way in which a normal distribution arises theoretically, the parameters can be thought of as determining the quantity L and the number and size of variations (u) operating. Once these are fixed, the distribution is fixed exactly.

DECIDING WHETHER DATA ARE NORMALLY DISTRIBUTED

When it is suspected that a standard distribution can be applied to a situation, it may be prudent to check that the variable does, approximately, follow the distribution in question. Since the normal distribution is constructed on the basis of a theoretical situation which is not likely to match exactly the real situation, the observations should be compared to what is theoretically expected. For example, do the following data have, approximately, a normal distribution? The data are the numbers of breakdowns per month of a manufacturing plant.

$$5\ 8\ 2\ 4\ 4\ 6\ 5\ 3\ 7\ 6\ 9\ 7\ 4\ 2\ 6\ 5\ 3\ 5\ 5\ 4$$

The data will be better handled if they are grouped together:

Value	0 1 2 3 4 5 6 7 8 9
Frequency	2 2 4 5 3 2 1 1

First calculate the parameters, the arithmetic mean and standard deviation.

Mean = 5.0

Standard deviation = 1.9

What is theoretically expected from a normal distribution with these parameters is that 68% of the observations will be between 5 ± 1.9, 95% between 5 ± 3.8, etc. The comparison is shown in table 4.2. Although it is not perfect, the match between observed and theoretical for just 20 observations is reasonably good, suggesting an approximate normal distribution. Judgement alone has been used to say that the match is acceptable. Statistical approaches to this question are available but are beyond the scope of this text. The method is, however,

Table 4.2   Observed and theoretical (percentages)

	Observed %	Theoretical %
Mean ± 1s: 3.1–6.9	60	68
Mean ± 2s: 1.2–8.8	95	95
Mean ± 3s: −0.7–10.7	100	99.7

based on the same principle. The observed data are compared with what is theoretically expected under an assumption of normality.

THE USE OF NORMAL CURVE TABLES

A factory produces a wide variety of tins for use with food products. A particular machine produces the lids for tins of instant coffee. The lid diameters are normally distributed with a mean of 10 cm and a standard deviation of 0.03 cm.

(a) What percentage of the lids have diameters in the range 9.97 cm to 10.03 cm?

Information met so far about areas under normal curves refers to values of the variable being measured in terms of 'standard deviations away from the mean'. The range 9.97 to 10.03 when in this form is:

$$10 - 0.03 \text{ to } 10 + 0.03$$

i.e. mean − 1s to mean + 1s

From the properties of a normal curve (Figure 4.6), this range includes 68% of the area under the curve or, in other words, 68% of the observations.
So 68% of the lids have diameters in the range 9.97 to 10.03 cm.

(b) Lids of diameter greater than 10.05 cm are too large to fit and must be discarded. If the machine produces 8000 lids per shift, how many are wasted?

Lids wasted have diameters greater than 10.05 Or, greater than mean + 1.67 standard deviations (since $(10.05 - 10.00)/0.03 = 1.67$)

The range is therefore no longer expressed as the mean + a whole number of standard deviations. For ranges such as these, a normal curve table, such as table 4.3, gives the necessary information. The table is used as follows:

Each number in the body of the table is the area (as a proportion of the whole area) under the curve from the mean to a point a given number (z) of standard deviations to the right of the mean (the shaded area in figure 4.8). In this example the area under the curve from the mean to the point 1.67 standard deviations to the right is wanted. In table 4.3, look down the left-hand column to find 1.6, then across the top row to find 0.07. The figure at the intersection is 0.4525 corresponding to 1.67 standard deviations from the mean.

The question asks for the probability of a diameter greater than 10.05 cm, i.e. for the area beyond 1.67 standard deviations. The area to the right of the mean as far as 1.67 standard deviations has just been found to be 0.4525; the entire area to the right of the mean is 0.5, since the distribution is symmetrical.
Therefore (see figure 4.9),

$$P \text{ (diameter greater than } 10.05) = 0.5 - 0.4525$$

$$= 0.0475$$

i.e. 4.75% of lids will have a diameter greater than 10.05. In a shift, 4.75% of 8000 = 380 will be unusable.

(c) Lids that are too small also have to be discarded. Diameters less than 9.93 cm do not provide a sufficiently airtight seal and are unusable. How many lids per shift have to be discarded because they are either too large or too small?

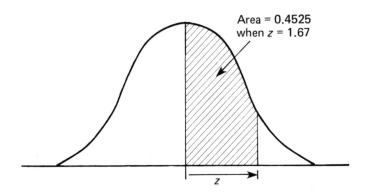

Area = 0.4525
when z = 1.67

Figure 4.8

Table 4.3   The normal distribution

z	0.00	.01	.02	.03	.04	.05	.06	.07	.08	.09
0.0	.0000	.0040	.0080	.0120	.0160	.0199	.0239	.0279	.0319	.0359
0.1	.0398	.0438	.0478	.0517	.0557	.0596	.0636	.0675	.0714	.0753
0.2	.0793	.0832	.0871	.0910	.0948	.0987	.1026	.1064	.1103	.1141
0.3	.1179	.1217	.1255	.1293	.1331	.1368	.1406	.1443	.1480	.1517
0.4	.1554	.1591	.1628	.1664	.1700	.1736	.1772	.1808	.1844	.1879
0.5	.1915	.1950	.1985	.2019	.2054	.2088	.2123	.2157	.2190	.2224
0.6	.2257	.2291	.2324	.2357	.2389	.2422	.2454	.2486	.2517	.2549
0.7	.2580	.2611	.2642	.2673	.2704	.2734	.2764	.2794	.2823	.2852
0.8	.2881	.2910	.2939	.2967	.2995	.3023	.3051	.3078	.3106	.3133
0.9	.3159	.3186	.3212	.3238	.3264	.3289	.3315	.3340	.3365	.3389
1.0	.3413	.3438	.3461	.3485	.3508	.3531	.3554	.3577	.3599	.3621
1.1	.3643	.3665	.3686	.3708	.3729	.3749	.3770	.3790	.3810	.3830
1.2	.3849	.3869	.3888	.3907	.3925	.3944	.3962	.3980	.3997	.4015
1.3	.4032	.4049	.4066	.4082	.4099	.4115	.4131	.4147	.4162	.4177
1.4	.4192	.4207	.4222	.4236	.4251	.4265	.4279	.4292	.4306	.4319
1.5	.4332	.4345	.4357	.4370	.4382	.4394	.4406	.4418	.4429	.4441
1.6	.4452	.4463	.4474	.4484	.4495	.4505	.4515	.4525	.4535	.4545
1.7	.4554	.4564	.4573	.4582	.4591	.4599	.4608	.4616	.4625	.4633
1.8	.4641	.4649	.4656	.4664	.4671	.4678	.4686	.4693	.4699	.4706
1.9	.4713	.4719	.4726	.4732	.4738	.4744	.4750	.4756	.4761	.4767
2.0	.4772	.4778	.4783	.4788	.4793	.4798	.4803	.4808	.4812	.4817
2.1	.4821	.4826	.4830	.4834	.4838	.4842	.4846	.4850	.4854	.4857
2.2	.4861	.4864	.4868	.4871	.4875	.4878	.4881	.4884	.4887	.4890
2.3	.4893	.4896	.4898	.4901	.4904	.4906	.4909	.4911	.4913	.4916
2.4	.4918	.4920	.4922	.4925	.4927	.4929	.4931	.4932	.4934	.4936
2.5	.4938	.4940	.4941	.4943	.4945	.4946	.4948	.4949	.4951	.4952
2.6	.4953	.4955	.4956	.4957	.4959	.4960	.4961	.4962	.4963	.4964
2.7	.4965	.4966	.4967	.4968	.4969	.4970	.4971	.4972	.4973	.4974
2.8	.4974	.4975	.4976	.4977	.4977	.4978	.4979	.4979	.4980	.4981
2.9	.4981	.4982	.4982	.4983	.4984	.4984	.4985	.4985	.4986	.4986
3.0	.4987	.4987	.4987	.4988	.4988	.4989	.4989	.4989	.4990	.4990

First, the percentage of diameters that are less than 9.93 has to be found. In terms of standard deviations this is, $z = (10 - 9.93)/0.03 = 2.33$ standard deviations to the left of the mean. Since the distribution is symmetrical, the table can equally well be used for

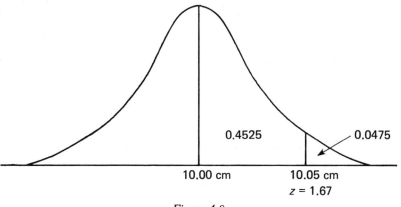

Figure 4.9

areas either side of the mean. As before, find 2.3 down the left column and 0.03 along the top row. At the intersection, the table gives a value of 0.4901.

$$P \text{ (diameter 9.93 to 10.0)} = 0.4901$$

$$P \text{ (diameter} < 9.93) \qquad = 0.5 - 0.4901$$

$$= 0.0099$$

i.e. 0.99% of the lids are too small.
In a shift, 0.99% of 8000 = 79 are too small.

In total, 79 + 380 lids are unusable = 459 per shift.

(d) Between what limits are 90% of the production?

How many standard deviations (z) will cover an area of 0.45 to the right of the mean and an area 0.45 to the left? See figure 4.10. Use the normal table 4.3 in reverse. In the body of the table, find 0.45. 0.4495 is in the row 1.6 and the column 0.04; 0.4505 is in the row 1.6 and the column 0.05. Approximately, then, the area 0.45 corresponds to 1.645 standard deviations.

90% of the production lies within ± 1.645 s of the mean

Or, within ± 1.645 x 0.03 cm of the mean = ±0.04935 cm. The range of diameters for 90% of the lids is 9.951 to 10.049 cm (approx.)

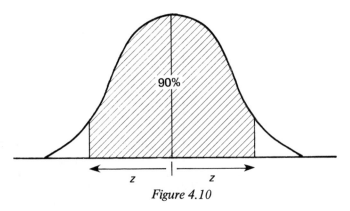

Figure 4.10

TECHNICAL NOTE

The mathematics of the normal distribution are exceedingly complicated. Fortunately the tables are easy to use and it is not usually necessary in applications to know the mathematical formula for the distribution. This is an example of the 'black box'.

Binomial distribution

After the normal distribution, the next most commonly found standard distribution is probably the binomial distribution.

CHARACTERISTICS

The binomial distribution is discrete (the values taken by the variable are distinct) giving rise to stepped shapes rather than a smooth curve. Figure 4.11 illustrates this, and also that the shape varies from right-

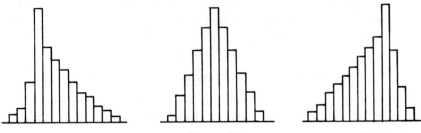

Figure 4.11

skewed through symmetrical to left-skewed depending upon the parameter values. The parameters of this distribution (which are not the mean and standard deviation) will be explained later.

SITUATIONS IN WHICH THE BINOMIAL OCCURS

Just as the normal distribution is constructed mathematically from a theoretical situation, so it is with the binomial. Before looking at the binomial situation, two definitions must be noted. First, a statistical *population* is the set of all possible values of a variable. Second, a *random sample* is one in which each element in the sample has an equal chance of being chosen. The situation is as follows:

The elements of a statistical population are of two types (hence the name of the distribution, bi-nomial). Each element must be of one but only one type. The proportion p of the population that is of the first type is known (and the proportion of the second type is therefore $1 - p$). A random sample of size n is taken. Because the sample is random, the number of elements of the first type it contains is not certain (it could be 0, 1, 2, 3, . . . or n).

If many samples were collected, a distribution would be gradually built up, the variable being the number of elements of the first type in each sample. From this theoretical situation the probabilities that the sample contains given numbers of elements of the first type can be calculated (but not easily, some mathematics and probability theory are required). This distribution is the binomial.

For example, a machine produces micro-chip circuits for use in children's toys. The circuits can be tested and found to be defective or OK. The machine has been designed to produce no more than 5% defective chips. A sample of 20 chips is collected. Assuming that overall exactly 5% of chips are defective, the binomial distribution probabilities give the likelihood that the sample contains 0, 1, 2 . . . or 20 defective chips. The probabilities can be found in a binomial distribution table (such as table 4.4). If several samples of size 20 are taken a distribution of the number of defectives per sample will emerge. The frequencies of the distribution should match the theoretically calculated probabilities. If they do not, then this may be an indication that the assumption of 5% defectives is not correct.

Actual situations in which the binomial is used include:

- Inspection schemes.
- Gallup polls (the 2-way split of the population results from agreement/disagreement with statements made by the pollster).
- Selling (the split is the outcome of a contact being a sale/no sale).

PARAMETERS

The parameters of a binomial distribution are the population proportion of one type (p) and the sample size (n). For different values of n and p the distribution shape varies (see figure 4.11). Fixing the values of n and p fixes the distribution exactly. The technical note below shows why this is so.

DECIDING WHETHER DATA FIT A BINOMIAL DISTRIBUTION

The theoretical situation on which the binomial distribution is defined is less abstract than for the normal. The definition of the normal is based on many assumptions (many small, additive, independent variations from a central value). It is difficult to know the extent to which the assumptions are met. The binomial assumptions are fewer. It is easier to predict, from knowledge of an actual situation, whether a binomial will apply. Even so, a check should be made. As with the normal, this is done by collecting observations and comparing them with what would be expected theoretically if the binomial applied. There is an example of this at the end of the chapter.

USING BINOMIAL DISTRIBUTION TABLES

A manufacturer has a contract with a supplier that no more than 5% of the supply of a particular component will be defective. The component is delivered in lorry-loads of 500. From each delivery a random sample of 20 is taken and inspected. If more than 3 of the 20 are defective the load is rejected. What is the probability of a load being rejected even though the supplier is sending an acceptable proportion of 5% defective?

The binomial distribution is applicable, since the population is split 2 ways and the samples are random. Table 4.4 is a binomial table for samples of size 20. The rows refer to the number of defective components in the sample; the columns refer to the proportion of defectives in the population of all components supplied (0.05 here).

Table 4.4   Binomial probabilities; sample size (n) = 20

					p					
r	0.05	.10	.15	.20	.25	.30	.35	.40	.45	.50
0	.3585	.1216	.0388	.0115	.0032	.0008	.0002	.0000	.0000	.0000
1	.3774	.2702	.1368	.0576	.0211	.0068	.0020	.0005	.0001	.0000
2	.1887	.2852	.2293	.1369	.0669	.0278	.0100	.0031	.0008	.0002
3	.0596	.1901	.2428	.2054	.1339	.0716	.0323	.0123	.0040	.0011
4	.0133	.0898	.1821	.2182	.1897	.1304	.0738	.0350	.0139	.0046
5	.0022	.0319	.1028	.1746	.2023	.1789	.1272	.0746	.0365	.0148
6	.0003	.0089	.0454	.1091	.1686	.1916	.1712	.1244	.0746	.0370
7	.0000	.0020	.0160	.0545	.1124	.1643	.1844	.1659	.1221	.0739
8	.0000	.0004	.0046	.0222	.0609	.1144	.1614	.1797	.1623	.1201
9	.0000	.0001	.0011	.0074	.0271	.0654	.1153	.1597	.1771	.1602
10	.0000	.0000	.0002	.0020	.0099	.0308	.0686	.1171	.1593	.1762
11	.0000	.0000	.0000	.0005	.0030	.0120	.0336	.0710	.1185	.1602
12	.0000	.0000	.0000	.0001	.0008	.0039	.0136	.0355	.0727	.1201
13	.0000	.0000	.0000	.0000	.0002	.0010	.0045	.0146	.0366	.0739
14	.0000	.0000	.0000	.0000	.0000	.0002	.0012	.0049	.0150	.0370
15	.0000	.0000	.0000	.0000	.0000	.0000	.0003	.0013	.0049	.0148
16	.0000	.0000	.0000	.0000	.0000	.0000	.0000	.0003	.0013	.0046
17	.0000	.0000	.0000	.0000	.0000	.0000	.0000	.0000	.0002	.0011
18	.0000	.0000	.0000	.0000	.0000	.0000	.0000	.0000	.0000	.0002
19	.0000	.0000	.0000	.0000	.0000	.0000	.0000	.0000	.0000	.0000
20	.0000	.0000	.0000	.0000	.0000	.0000	.0000	.0000	.0000	.0000

Table 4.5   Component example

P (0 defectives) = 35.8%
P (1 defective) = 37.7%
P (2 defectives) = 18.9%
P (3 defectives) = 6.0%
P (more than 3)= 1.6%

Total = 100.0%

Table 4.5 is based on the column headed 0.05 in table 4.4. The manufacturer rejects loads when there are 3 or more defectives in the sample. From table 4.5 the probability of a load being accepted even though the supplier is sending an acceptable proportion of defectives is 1.6%.

TECHNICAL NOTE

Binomial distribution tables are lengthy. Table 4.4 relates to just one sample size (20). For anyone in possession of a slightly mathematical mind and, more especially, a calculator, it is often easier to work out binomial probabilities directly. The derivation of the formula is left in its black box.

The mathematical formula by which binomial probabilities are calculated is:

$$P \text{ (}r \text{ of type 1 in sample)} = nCr \times p^r \times (1 - p)^{n-r}$$

The symbols are:

r = number in sample that are of type 1

n = sample size

p = proportion of type 1 in population

$1 - p$ = proportion of type 2 in population

nCr is mathematical shorthand for a 'combination'
$nCr = n!/(r! \times (n - r)!)$

$n!$ (said n factorial) $= n \times (n - 1) \times (n - 2) \times \ldots \times 2 \times 1$
e.g. $4! = 4 \times 3 \times 2 \times 1 = 24$

The following example shows how this formula can be used.

If 40% of the USA electorate are Republican voters, what is the probability that a randomly assembled group of 3 will contain 2 Republicans? Using the binomial formula,

P (2 Republicans in group of 3)

$$= P \text{ (}r = 2)$$
$$= nC2 \times p^2 \times (1 - p)^{n-2}$$
$$= 3C2 \times p^2 \times (1 - p)^{3-2}$$
$$= 3C2 \times 0.4^2 \times 0.6$$
$$= 3!/(2! \times 1!) \times 0.16 \times 0.6$$
$$= 3 \times 0.16 \times 0.6$$
$$= 0.288$$

There is a 28.8% chance that a group of 3 will contain 2 Republican voters.

Making similar calculations for $r = 0$, 1 and 3, the binomial distribution of figure 4.12 is obtained. The results may be checked using table 4.6, which is the binomial table for sample size 3 (and other sizes from 1 to 7).

Poisson distribution

The Poisson distribution is a standard distribution closely linked to the binomial.

CHARACTERISTICS

The Poisson is a discrete distribution. Its shape varies from right skewed to almost symmetrical as in figure 4.13.

SITUATIONS IN WHICH THE POISSON OCCURS

The Poisson distribution may be thought of as similar to the binomial but with an infinite sample size. The split of the population into two types is into the occurrence of events and the 'non-occurrence of events'. The distribution allows the probabilities that given numbers of events occur within the sample to be calculated. Compare this with the binomial whereby the probabilities that the sample contains given numbers of elements of one type are calculated. The mathematics of the Poisson distribution are based on those of the binomial, but changed to allow for an infinite sample size.

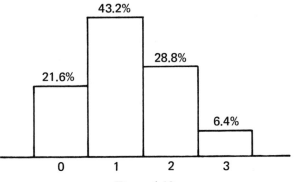

Figure 4.12

Table 4.6   Binomial table for sample sizes (n) 1 to 7

						p					
n	r	0.05	.10	.15	.20	.25	.30	.35	.40	.45	.50
1	0	.9500	.9000	.8500	.8000	.7500	.7000	.6500	.6000	.5500	.5000
	1	.0500	.1000	.1500	.2000	.2500	.3000	.3500	.4000	.4500	.5000
2	0	.9025	.8100	.7225	.6400	.5625	.4900	.4225	.3600	.3025	.2500
	1	.0950	.1800	.2550	.3200	.3750	.4200	.4550	.4800	.4950	.5000
	2	.0025	.0100	.0225	.0400	.0625	.0900	.1225	.1600	.2025	.2500
3	0	.8574	.7290	.6141	.5120	.4219	.3430	.2746	.2160	.1664	.1250
	1	.1354	.2430	.3251	.3840	.4219	.4410	.4436	.4320	.4034	.3750
	2	.0071	.0270	.0574	.0960	.1406	.1890	.2389	.2880	.3341	.3750
	3	.0001	.0010	.0034	.0080	.0156	.0270	.0429	.0640	.0911	.1250
4	0	.8145	.6561	.5220	.4096	.3164	.2401	.1785	.1296	.0915	.0625
	1	.1715	.2916	.3685	.4096	.4219	.4116	.3845	.3456	.2995	.2500
	2	.0135	.0486	.0975	.1536	.2109	.2646	.3105	.3456	.3675	.3750
	3	.0005	.0036	.0115	.0256	.0469	.0756	.1115	.1536	.2005	.2500
	4	.0000	.0001	.0005	.0016	.0039	.0081	.0150	.0256	.0410	.0625
5	0	.7738	.5905	.4437	.3277	.2373	.1681	.1160	.0778	.0503	.0312
	1	.2036	.3280	.3915	.4096	.3955	.3602	.3124	.2592	.2059	.1562
	2	.0214	.0729	.1382	.2048	.2637	.3087	.3364	.3456	.3369	.3125
	3	.0011	.0081	.0244	.0512	.0879	.1323	.1811	.2304	.2757	.3125
	4	.0000	.0004	.0022	.0064	.0146	.0284	.0488	.0768	.1128	.1562
	5	.0000	.0000	.0001	.0003	.0010	.0024	.0053	.0102	.0185	.0312
6	0	.7351	.5314	.3771	.2621	.1780	.1176	.0754	.0467	.0277	.0156
	1	.2321	.3543	.3993	.3932	.3560	.3025	.2437	.1866	.1359	.0938
	2	.0305	.0984	.1762	.2458	.2966	.3241	.3280	.3110	.2780	.2344
	3	.0021	.0146	.0415	.0819	.1318	.1852	.2355	.2765	.3032	.3125
	4	.0001	.0012	.0055	.0154	.0330	.0595	.0951	.1382	.1861	.2344
	5	.0000	.0001	.0004	.0015	.0044	.0102	.0205	.0369	.0609	.0938
	6	.0000	.0000	.0000	.0001	.0002	.0007	.0018	.0041	.0083	.0516
7	0	.6983	.4783	.3206	.2097	.1335	.0824	.0490	.0280	.0152	.0078
	1	.2573	.3720	.3960	.3670	.3115	.2471	.1848	.1306	.0872	.0547
	2	.0406	.1240	.2097	.2753	.3115	.3177	.2985	.2613	.2140	.1641
	3	.0036	.0230	.0617	.1147	.1730	.2269	.2679	.2903	.2918	.2734
	4	.0002	.0026	.0109	.0287	.0577	.0972	.1442	.1935	.2388	.2734
	5	.0009	.0002	.0012	.0043	.0115	.0250	.0466	.0774	.1172	.1641
	6	.0000	.0000	.0001	.0004	.0013	.0036	.0084	.0172	.0320	.0547
	7	.0000	.0000	.0000	.0000	.0001	.0002	.0006	.0016	.0037	.0078

Table 4.7   Poisson distribution table

	λ									
r	0.005	.01	.02	.03	.04	.05	.06	.07	.08	.09
0	.9950	.9900	.9802	.9704	.9608	.9512	.9418	.9324	.9231	.9139
1	.0050	.0099	.0196	.0291	.0384	.0476	.0565	.0653	.0738	.0323
2	.0000	.0000	.0002	.0004	.0008	.0012	.0017	.0023	.0030	.0037
3	.0000	.0000	.0000	.0000	.0000	.0000	.0000	.0001	.0001	.0001

	λ									
r	0.1	0.2	0.3	0.4	0.5	0.6	0.7	0.8	0.9	1.0
0	.9048	.8187	.7408	.6703	.6065	.5488	.4966	.4493	.4066	.3679
1	.0905	.1637	.2222	.2681	.3033	.3293	.3476	.3595	.3659	.3679
2	.0045	.0164	.0333	.0536	.0758	.0988	.1217	.1433	.1647	.1839
3	.0002	.0011	.0033	.0072	.0126	.0198	.0284	.0383	.0494	.0613
4	.0000	.0001	.0002	.0007	.0016	.0030	.0050	.0077	.0111	.0153
5	.0000	.0000	.0000	.0001	.0002	.0004	.0007	.0012	.0020	.0031
6	.0000	.0000	.0000	.0000	.0000	.0000	.0001	.0002	.0003	.0005
7	.0000	.0000	.0000	.0000	.0000	.0000	.0000	.0000	.0000	.0001

	λ									
r	1.1	1.2	1.3	1.4	1.5	1.6	1.7	1.8	1.9	2.0
0	.3329	.3012	.2725	.2466	.2231	.2019	.1827	.1653	.1496	.1353
1	.3662	.3614	.3543	.3452	.3347	.3230	.3106	.2975	.2842	.2707
2	.2014	.2169	.2303	.2417	.2510	.2584	.2640	.2678	.2700	.2707
3	.0738	.0867	.0998	.1128	.1255	.1378	.1496	.1607	.1710	.1804
4	.0203	.0260	.0324	.0395	.0471	.0551	.0636	.0723	.0812	.0902
5	.0045	.0062	.0084	.0111	.0141	.0176	.0216	.0260	.0309	.0361
6	.0003	.0012	.0018	.0026	.0035	.0047	.0061	.0078	.0098	.0120
7	.0001	.0002	.0003	.0005	.0008	.0011	.0015	.0020	.0027	.0034
8	.0000	.0000	.0001	.0001	.0001	.0002	.0003	.0005	.0006	.0009
9	.0000	.0000	.0000	.0000	.0000	.0000	.0001	.0001	.0001	.0002

	λ									
r	2.1	2.2	2.3	2.4	2.5	2.6	2.7	2.8	2.9	3.0
0	.1225	.1103	.1003	.0907	.0821	.0743	.0672	.0608	.0550	.0498
1	.2572	.2438	.2306	.2177	.2052	.1931	.1815	.1703	.1596	.1494
2	.2700	.2681	.2652	.2613	.2565	.2510	.2450	.2384	.2314	.2240
3	.1890	.1966	.2033	.2090	.2138	.2176	.2205	.2225	.2237	.2240
4	.0992	.1082	.1169	.1254	.1336	.1414	.1483	.1557	.1622	.1680
5	.0417	.0476	.0538	.0602	.0668	.0735	.0804	.0872	.0940	.1008
6	.0146	.0174	.0206	.0241	.0278	.0319	.0362	.0407	.0455	.0504
7	.0044	.0055	.0068	.0083	.0099	.0118	.0139	.0163	.0188	.0216
8	.0011	.0015	.0019	.0025	.0031	.0038	.0047	.0057	.0068	.0081
9	.0003	.0004	.0005	.0007	.0009	.0011	.0014	.0018	.0022	.0027
10	.0001	.0001	.0001	.0002	.0002	.0003	.0004	.0005	.0006	.0008
11	.0000	.0000	.0000	.0000	.0000	.0001	.0001	.0001	.0002	.0002
12	.0000	.0000	.0000	.0000	.0000	.0000	.0000	.0000	.0000	.0001

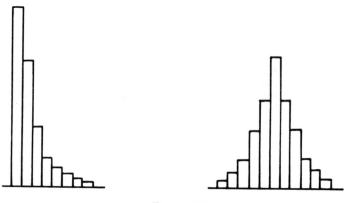

Figure 4.13

A typical use is the arrival of telephone calls at a switchboard. The sample is a period of time. The elements of the sample are the arrival or non-arrival of calls. Since, potentially, an infinite number of calls could arrive during the time-period, the sample size is in effect infinite. The number of calls that do arrive can be counted, but there remains an uncountable number of calls that did not arrive. Given an average arrival rate of calls (the average number of calls per time-period), the probability that any number of calls will arrive can be calculated. For instance, knowing that 10 calls per minute arrive on average, the Poisson gives the probabilities that in any minute 0, 1, 2, 3, 4 . . . calls will arrive. The probabilities can be found in Poisson distribution tables (see table 4.7).

Other situations in which the Poisson occurs are:

- Flaws in telegraph cable (there are only a finite number of flaws in a given length of cable, but an infinite number could potentially have occurred but did not).
- Mechanical breakdowns of machinery, cars, etc.
- Road accidents.

PARAMETERS

The Poisson has just one parameter, the average occurrence of events (the average number of calls per minute in the above example). Once this parameter (written as λ, said lambda) is known, the shape of the distribution is fixed exactly.

DECIDING WHETHER DATA FIT A POISSON DISTRIBUTION

As with the normal and binomial distributions, the theoretical basis of the Poisson can never quite be matched in practice. There are two tests to apply to check whether the Poisson is applicable. First, the actual situation is compared with that on which the distribution is based. Second, some observed data are compared with what is theoretically expected. An example of this is given at the end of the chapter.

USING THE POISSON DISTRIBUTION TABLES

A company receives, on average, 2 telephone calls per minute. What capacity switchboard (in terms of calls per minute) should be installed so that it can cope with incoming calls 95% of the time?

Assuming the average number of calls/minute does not vary during the switchboard's working day, the Poisson can be used. The assumption is not restrictive since the day can be divided into sections reflecting different levels of telephone activity. The assumption is important since the Poisson is based (like the binomial) on taking a random sample, not a sample from a section of the day when the parameter value may be different. If 2 calls/minute is the average for the whole day, then in busy periods the switchboard may not cope as well as the calculations suggest.

Table 4.7 shows Poisson probabilities for varying parameter values. Each row of the table refers to a particular number of events occurring in the sample; each column refers to a particular parameter value. For this example the parameter is 2 (an average of 2 calls per minute). The column associated with a parameter value of 2 gives the probabilities in table 4.8, which shows that there will be 4 or fewer calls 94.7% of the time. A switchboard of capacity 4 will therefore be able to handle incoming calls (approximately) 95% of the time.

Table 4.8

No. calls	Probability (%)
0	13.5
1	27.1
2	27.1
3	18.0
4	9.0
5	3.6
6	1.2
7 or more	0.5
	100.0

The values 13.5, 27.1, 27.1, 18.0, 9.0 are braced together totalling 94.7

TECHNICAL NOTE

The Poisson distribution is derived from the binomial. Starting with the binomial formula for probabilities:

$$P \text{ (}r \text{ of type 1)} = nCr \times p^r \times (1 - p)^{n-r}$$

If n increases indefinitely while np (the average number of the first type per sample) remains at some constant level, then the formula becomes (after some not inconsiderable mathematics):

$$P \text{ (}r \text{ events)} = e^{-\lambda} \cdot \lambda^r / r!$$

λ is the parameter of the distribution, the average number of events per sample; e is a constant, equal to 2.718 . . . which just happens to occur in certain mathematical situations (cf. the way π, equal to 3.141 . . . , just happens to occur as the ratio between the circumference and diameter of a circle).

The formula is the means of calculating the probabilities in table 4.7. It could have been applied to the switchboard example above.

The average number of calls per minute is 2, i.e. λ is 2. From tables or a calculator, $e^{-2.0} = 0.135$, thus

$$P \text{ (0 calls)} = 0.135 \times 1/1 = 0.135 \quad (0! = 1, \text{ by definition})$$

$$P \text{ (1 call)} = 0.135 \times 2/1 = 0.27$$

and so on, replicating Table 4.8

As with the binomial it is sometimes, with the aid of a calculator, easier to use the formula than the lengthy tables; especially when it is noted from the Poisson formula that

$$P(r + 1) = P(r) \text{ times } \lambda/(r + 1)$$

e.g. $P(r = 1) = P(r = 0) \times \lambda$

$P(r = 2) = P(r = 1) \times \lambda/2$

$P(r = 3) = P(r = 2) \times \lambda/3$, etc.

Approximations to the binomial

Of the three standard distributions encountered, the binomial is the most difficult to use because of its complicated formula and lengthy probability tables. The normal and Poisson are easier to use, the former because of the simple table and the latter because of its relatively straightforward formula. For certain parameter values the shape of the binomial is similar to the normal; for others it is similar to the Poisson. In these circumstances, purely for ease of use, the binomial is approximated by the normal and Poisson distributions respectively.

USING THE NORMAL TO APPROXIMATE THE BINOMIAL

The binomial is roughly symmetrical when p is not close to 0 or when n is large. As a rule of thumb, the binomial can be approximated by the normal provided np and $n(1 - p)$ both exceed 5. For the binomial, using some more 'black box' mathematics:

$$\text{Mean} = np$$

$$\text{Standard deviation} = \sqrt{np(1 - p)}$$

These are the parameters of the normal distribution which can then be used, as if the distribution were normal.

EXAMPLE

If, on average, 20% of firms respond to industrial questionnaires, what is the probability of fewer than 50 responses when 300 questionnaires are sent out?

The situation is one to which the binomial applies. Firms either respond or do not respond, giving the 2-way split of the population. A sample of 300 is being taken from this population. To answer the question using the binomial formula, the following would have to be calculated or obtained from tables:

$$P \text{ (0 responses)} = 300C0 \times 0.2^0 \times 0.8^{300}$$
$$P \text{ (1 response)} = 300C1 \times 0.2^1 \times 0.8^{299}$$

.

.

.

$$P \text{ (49 responses)} = 300C49 \times 0.2^{49} \times 0.8^{251}$$

These 50 probabilities would then have to be added. This is a daunting task, even for mathematical masochists. However, since $np = 60 (= 300 \times 0.2)$ and $n(1 - p) = 240$ are both greater than 5, the normal approximation can be used, with parameters:

$$\text{Mean} = np = 60$$
$$\text{Standard deviation} = \sqrt{np(1 - p)}$$
$$= \sqrt{300 \times 0.2 \times 0.8}$$
$$= 7, \text{approx.}$$

A slight difficulty arises in that a discrete distribution is being approximated by a continuous one. To make the approximation it is necessary to pretend that the variable (the number of responses) is continuous. For example, 60 responses represents the interval between 59.5 and 60.5 and so on.

Consequently, to answer the question,

$$P \text{ } (r < 49.5)$$

must be found.

49.5 is 10.5 from the mean

or, $z = 10.5/7 = 1.5$ standard deviations from the mean.

Using the normal table 4.3:

$$P \text{ } (z < 1.5) = 0.5 - 0.4332$$
$$= 0.0668$$
$$P \text{ (fewer than 50 replies)} = 6.68\%$$

USING THE POISSON TO APPROXIMATE THE BINOMIAL

Because of the way the distributions are linked, the binomial can be approximated by the Poisson when the sample is large and the proportion p is small. As a rule of thumb, the approximation will give good results if $n \geqslant 20$ and $p \leqslant 0.05$. The parameter of the approximating Poisson is easily found since λ is defined as being equal to the mean of the binomial, np (but again some mathematics is needed to prove this).

EXAMPLE

15,000 cars per day use the traffic tunnel at a major international airport. The chance of any 1 car breaking down in the tunnel is 0.00003 (from historical records). Should 3 cars break down it is unlikely that emergency services will be able to deal with the situation and traffic will come to a standstill. What is the probability that there will be no more than 2 breakdowns in the tunnel?

The binomial applies to this situation. Using the formula, $n = 15,000$ and $p = 0.00003$.

P (0 breakdowns) $= P\ (r = 0) = 15,000C0 \times (0.00003)^0 \times (0.99997)^{15,000}$

P (1 breakdown) $= P\ (r = 1) = 15,000C1 \times (0.00003)^1 \times (0.99997)^{14,999}$

P (2 breakdowns) $= P\ (r = 2) = 15,000C2 \times (0.00003)^2 \times (0.99997)^{14,998}$

$P\ (r = 0) + P\ (r = 1) + P\ (r = 2)$ is the probability to be found.

The calculations are possible but, once again, daunting. Since $n \geqslant 20$ and $p \leqslant 0.05$, the Poisson approximation can be used.

$$\lambda = np = 15,000 \times 0.00003$$

$$= 0.45$$

From the Poisson formula: $(P\ (r) = e^{-\lambda} \times \lambda^r / r!)$

$$P\ (r = 0) = e^{-0.45} = 0.64$$

$$P\ (r = 1) = 0.64 \times 0.45 = 0.29$$

$$P\ (r = 2) = 0.29 \times 0.45/2 = 0.06$$

Therefore, P (at most 2 breakdowns) $= 0.64 + 0.29 + 0.06$
$$= 0.99$$

Worked examples

1. MARYLEBONE BUSINESS ACADEMY

In a large class in business statistics, the final examination grades have a mean of 67.4 and a standard deviation of 12. Assuming that the distribution of these grades (all whole numbers) is normal, find:

(a) the percentage of grades that should exceed 85;
(b) the percentage less than 40;
(c) the number of failures (pass = 50%) in a class of 180;
(d) the lowest distinction mark if the highest 8% of grades are to be regarded as distinctions.

(a) a grade of 85 is 17.6 away from the mean (85−67.4 = 17.6). In terms of standard deviations this is:

$$z = 17.6/12$$

$$= 1.47$$

Referring to the normal curve table 4.3, for $z = 1.47$ the shaded area A is 0.4292 (see figure 4.14).

$$\text{Area B} = 0.5 - 0.4292$$

$$= 0.0708$$

Therefore 7.1% of students should exceed a mark of 85.

(b) 40 is 27.4 away from the mean

$$z = -27.4/12$$

$$= -2.28$$

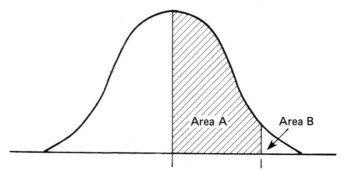

Figure 4.14

Referring to the normal curve table, the area between the mean and $z = -2.28$ is 0.4887. Therefore $0.5 - 0.4887 = 0.0113$, or 1.1% of students should get less than 40.

(c) 50 is 17.4 away from the mean

$$z = -17.4/12$$

$$= -1.45$$

Referring to the normal curve table, the area between the mean and $z = -1.45$ is 0.4265.

The proportion of students failing $= 0.5 - 0.4265$

$$= 0.0735$$

In a class of 180, this means $0.0735 \times 180 = 13$ students.

(d) If 8% of students are awarded distinctions, then the lowest distinction mark corresponds to the z value associated with the shaded area (figure 4.15) being equal to 0.42. From the table, this z value is 1.405. The lowest distinction mark is $67.4 + 1.405 \times 12 = 84$.

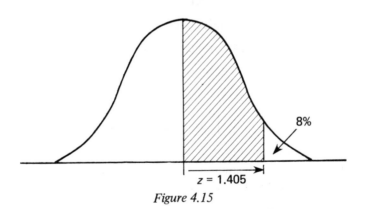

Figure 4.15

2. P. LUGG

The company manufactures electrical components. The quality control scheme for a particular type of component consists of taking random samples of 5 components at regular intervals. The numbers of defective components are counted and charted. Out of 100 such samples, in 60 cases no defectives were found, in 32 cases 1 defective, and in the remaining cases 2 defectives.

(a) Are these results consistent with the view that the process is operating with an average 10% of defectives?

(b) What reservations have you about your conclusion?

The population from which samples of size 5 are being taken is the entire production of these components. The population is split into 2 types, defective and non-defective. It is assumed that the process is operating with 10% defectives. Thus:

Binomial distribution applies with $p = 0.10$

$$n = 5$$

From the binomial table 4.6:

P (0 defectives) = 0.5905

P (1 defective) = 0.3280

P (2 defectives) = 0.0729

P (3 defectives) = 0.0081

P (4 defectives) = 0.0004

P (5 defectives) = 0

To see whether the results are consistent with an overall 10% defective rate, the observed results from the 100 samples are compared with the theoretically expected results calculated above.

Number of defectives	0	1	2	3	4	5
Observed no. of samples	60	32	8	0	0	0
Theoretical no. of samples	59	33	7	1	0	0

There is a good correspondence. The results are consistent with a process defective rate of 10%.

Reservations surrounding this conclusion depend upon whether the samples were taken at random. If the samples containing one or more defectives were clustered together, perhaps at the start or end of a shift, this might indicate that the true process defective rate were less than 10% but that there were particular problems in starting up and closing down the machines.

If the samples containing one or more defectives were mostly towards the end of the time period during which the samples were

collected, then this might indicate that the process used to have a
defective rate less than 10% but had deteriorated.

3. TITAN INSURANCE COMPANY

Merchant ships are subject to accident because of bad weather, running aground, ice, breakdowns, etc. For a certain type of vessel the
level of insurance premium to cover against these risks is to be decided.
To help with this decision, information on the accidents occurring to
100 vessels over a 400-day period has been collected:

No. accidents in 400 days (r)	0	1	2	3	4	5	6
No. vessels with r accidents	24	35	26	10	3	1	1

(a) Does this pattern of accidents follow a Poisson distribution?
(b) What further data or analyses are required to strengthen the
conclusion?

The situation looks to be Poisson. The sample is the 400-day period.
The events are the accidents. It is possible to count how many accidents have taken place, but not how many accidents might have taken
place but did not. There is thus a good case for supposing that the
Poisson will apply.

To test this, the observed results from 100 ships must be compared
with what would be expected theoretically under the assumption of
a Poisson distribution. In order to use the Poisson table 4.7, the
parameter of the distribution (the average number of accidents per
400 days) has to be calculated.

$$= (35 \times 1) + (26 \times 2) + (10 \times 3) + (3 \times 4) + (1 \times 5) + (1 \times 6)/100$$

$$= 140/100$$

$$= 1.4$$

From the column headed 1.4 in table 4.7 the theoretical frequencies
of accidents can be found.

No. accidents (r)	0	1	2	3	4	5	6+
Observed	24	35	26	10	3	1	1
Theoretical Poisson	25	35	24	11	4	1	0

The two criteria for judging whether a Poisson distribution fits the
data both suggest that it does. First, there is a good a priori case for

supposing the situation is Poisson. Second, the observed data are very close to what is anticipated theoretically.

Reservations about the conclusion are principally to do with the accidents being random or not. It may be that certain routes/types of ship/master mariners/times of the year are more prone to accident. If so, the average accident rate differs from one part of the population to another and a uniform value covering the whole population should not be used. In this case it may be necessary to treat each section of the population separately.

Another problem may be that the sample is not representative of the population, for example, not all routes may be included. If the insurance is to cover all routes then so should the sample data.

Final comments

A standard distribution is mathematically derived from a theoretical situation. If an actual situation matches (approximately) the theoretical then the standard distribution can be used both to describe and analyse the situation. As a result fewer data need be collected. This chapter has been concerned with three standard distributions: the normal, the binomial and the Poisson. For each the following have been described:

- its characteristics;
- the situations in which it can be used;
- its parameters;
- how to decide whether an actual situation matches the theoretical situation on which the distribution was based;
- an example of how it might be used.

The mathematics of the distributions have been hinted at, but not pursued rigorously. The underlying formulae require a relatively high level of mathematical and statistical knowledge. Fortunately such detail is not necessary for the effective use of the distributions. Furthermore, the role of the manager is unlikely to be that of a practitioner of statistics, rather he will have to supervise the use of statistical methods in his organization. It is therefore the central concepts of the distributions, not the mathematical detail, that are important. As was shown in the worked examples, the principal non-mathematical errors in the use of distributions is to base the analysis on non-random samples.

The distributions that have been the subject of this chapter are just three of the many that are available. The principles behind their use are the same, but each is associated with a different situation. To look at other standard distributions and to look at them more deeply goes beyond what a manager will find helpful to know and into the domain of the statistical expert.

Further Examples

1. C. HOUSEMAN INTERIOR PRODUCTS

One of the company's products is a range of bathroom tiles. The production process has an average 0.6% of tiles defective. The tiles are sold in packages containing 100 tiles. What standard distribution would apply to the variation in the number of defective tiles per package (and could thus be used as the basis for a quality control scheme)?

2. H. A. ROD STORES

This departmental store is monitoring the use of monthly credit accounts. A study shows that the average monthly expenditure of its regular account users is normally distributed with mean £320 and standard deviation £120. The customers are classified into 4 groups according to expenditure:

> Group 1 spends less than £200
> Group 2 spends £200 but less than £300
> Group 3 spends £300 but less than £400
> Group 4 spends £400 or more

What percentage of customers would you expect to fall within each group?

3. ORACLE OPINION POLL SERVICES

Random samples of 20 voters in the USA are asked if they would vote Republican were there an election tomorrow. Forty per cent of the total electorate are Republican voters. The number of Republican voters in the samples will vary, but within what range could you expect the number of Republican voters to be for 98% of such samples?

4. WESTERN AUSTRALIA INTERNAL TELECOMMUNICATIONS

On average the telegraph cable used by the company has 2.3 flaws per 5-mile stretch. What is the probability that the link between two small towns 5 miles apart will have at most one flaw?

ANSWERS

1. C. Houseman Interior Products

The situation is one to which the binomial is applicable, but since p is low (= 0.006) and n is large (= 100), the Poisson would be used in practice.

2. H. A. Rods Stores

Group 1: 15.9%
Group 2: 27.4%
Group 3: 31.6%
Group 4: 25.1%

3. Oracle Opinion Poll Services

It would be expected that 98% (to the nearest whole number) of samples would have from 4 to 13 Republican voters.

4. Western Australia Internal Telecommunications

The probability of at most one flaw is 33%.

CHAPTER 5

Statistical inference

By the end of the chapter the reader should understand the basic ideas behind statistical inference, without necessarily having sufficient expertise to be a practitioner. Statistical inference is the methods by which data from a single sample can be turned into more general information. It has two main parts. Estimation is concerned with making predictions and specifying their accuracy; significance tests are concerned with distinguishing between a result arising by chance and one arising from other factors.

Statistical inference is the use of sample data to predict, or infer, further pieces of information about the population from which the sample came. Inference is a collection of methods by which knowledge of a sample can be turned into knowledge about the population. Recall that the statistical population of a variable is the set of all possible values of the variable. One important form of inference is *estimation* where the sample is the basis for predicting the values of population parameters; a second is *significance tests*, which is a procedure for deciding whether sample evidence supports or rejects a hypothesis (a statement about a population).

The theoretical background is statistical distributions, especially the normal distribution. However, to make a fuller use of inferential methods some additional theory is required. This will be discussed before moving on to consider the processes of inference. First some examples of the applications of inference are given.

Applications of statistical inference

MARKET RESEARCH

Much of market research is based on investigating a sample of customers and then generalizing the results to the whole potential market. For example, a manufacturer of razor blades was re-designing the packaging of his product. The company wanted to know what proportion of the purchases of razor blades were made by women. The market research department looked at a random sample of 500 purchases made at several different outlets: 145 of the purchases (29%) were made by women. Statistical inference allowed the conclusion to be drawn that between 27% and 31% of all razor blade purchases are made by women.

MEDICINE

When a new medical treatment is developed it is both impossible and undesirable to give it to all suffering from the complaint. Usually it is given to a sample of sufferers and their progress compared to that of patients not given the treatment. A significance test makes it possible to say whether the new treatment is beneficial. For example, a new process for speeding the healing of pulled muscles had been developed. Hospital records showed that in the recent past healing had taken an average of 12 days with some variation on either side of the average. The new process was applied to a sample of 30 people with pulled muscles. On average healing took 10.5 days. The difference could be because the new treatment is better, or it could have arisen purely by chance since healing time does vary from person to person. A significance test is a method of distinguishing between the two causes. In this case the conclusion was that the evidence was not strong enough to suppose that the new treatment was better.

Confidence levels

A sample is merely a selection of data. More (the rest of the population) remain uncollected. Although drawn at random, a sample could still be unrepresentative of the population. Conclusions drawn from it may be wrong. To allow for this possibility, inferences are not stated with complete certainty. Conclusions are drawn such as: 'It is predicted, with 95% confidence, that the population mean percentage is in the range 27%–31%' (the razor blade example). It is not stated that the population mean is definitely in the range 27%–31%. The sample on which the prediction was based may have been unrepresentative. It is not known whether this was the case, but if a bet were to be made, there would be 19 chances in 20 that the mean was in the range given. This is the meaning of the '95% confidence'. A confidence level attached to a statement is the probability that the statement is true. All inferences are made at some level of confidence. In the medical example, the conclusion that the new process was no better was stated 'with 95% confidence', meaning that there was a 5% chance that the statement was wrong. By convention, 95% is the usual confidence level. The method of deriving confidence levels depends upon the next piece of theory.

Sampling distribution of the mean

Inference is accomplished more easily and more accurately with the help of a further distribution, that of the mean of a sample. Why this is so should become clearer later in the chapter. Imagine a series of samples all of the same size being taken at random from a population. If the mean of each sample were calculated, then this value would differ from sample to sample, purely by chance because the samples were chosen at random. It has, therefore, a distribution. It is referred to as the sampling distribution of the mean. The distribution has particular characteristics.

Figure 5.1(a) shows a variable which has a normal distribution. One can think of the variable as being the actual lengths of machine-produced rods of nominal length 100 cm (or any of the previous examples of normally distributed variables). A random sample of rods is taken and the arithmetic mean of the rods is calculated and recorded. A second sample of the same size is taken; and a third; and a fourth . . . Gradually a distribution of the sample means is

(a)   (b)

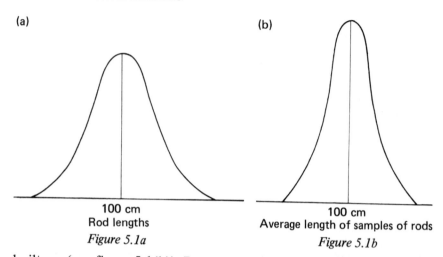

100 cm	100 cm
Rod lengths	Average length of samples of rods
Figure 5.1a	*Figure 5.1b*

built up (see figure 5.1(b)). Providing a very large number of samples are taken, the new distribution will be normal, have the same mean as the original, but be narrower. It is intuitively reasonable that the sampling distribution should have these characteristics. The narrowness occurs because of the tendency for extreme (very long or very short) rods to be 'averaged out' by the rods of more moderate length in the sample. The extent of the narrowing is measured by the distribution having a smaller standard deviation equal to:

Standard deviation of individual distribution$/\sqrt{\text{Sample size}}$

To prove that this and the other properties are true requires more than intuition. It requires some 'black box' mathematics. The relationship between the sampling distribution of the mean and the individual distribution from which the samples were drawn can be summarized:

Individual distribution	*Sampling distribution*
Normal	Normal
Mean = $\bar{x}$	Same mean = $\bar{x}$
Standard deviation = s	Standard deviation = $s/\sqrt{\text{Sample size}}$

If the individual distribution is not normal the outcome of taking samples is more surprising. Figure 5.2(a) shows a non-normal distribution. It could be any non-normal distribution. For example, the number of copies of a local weekly newspaper read in a year by the inhabitants of the small town that it serves, might have a distribution as in figure 5.2(a). A random sample of at least 30 inhabitants is taken. The arithmetic mean of the number of copies read by the

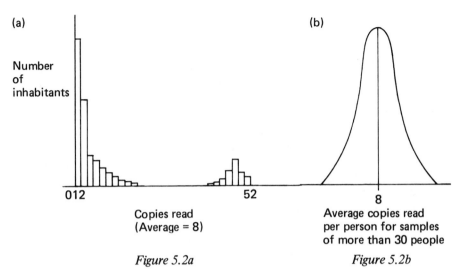

Figure 5.2a   *Figure 5.2b*

sample is calculated. More samples of the same size are taken. Eventually the distribution will be as shown in figure 5.2(b). The distribution is normal. The similarities and differences between the original individual distribution and the sampling distribution of the mean can be summarized:

Individual distribution	*Sampling distribution*
Non-normal	Normal
Mean = $\bar{x}$	Same mean = $\bar{x}$
Standard deviation = s	Standard deviation = $s/\sqrt{\text{Sample size}}$

Provided the sample size is greater than 30, the sampling distribution of the mean will be approximately normal whatever the shape of the distribution from which the samples were taken. The 30 is a rule of thumb. If the individual distribution is at all similar to a normal then a sample size of fewer than 30 will be enough to make the sampling distribution of the mean normal. For a slightly skewed distribution, a sample size of just 4 or 5 may well be sufficient. The links between the individual distribution can be established by observation (taking and recording samples) or mathematically.

The theory underlying this normalization property is a consequence of the Central Limit Thoerem. In view of the mathematics involved, only the results have been presented. The great benefit of these ideas is that even though the distribution of a variable is *unknown*, by taking samples the distribution of the sample mean is *known*, and some analysis can be done.

EXAMPLE

A manufacturing organization has a workforce of several thousands. Sickness records show that the average number of days each employee is off sick is 12, with a standard deviation of 5. If random samples of 100 employees were taken and the average number of days sickness per employee calculated for each sample, what would be the distribution of these sample averages? What would be its parameters?

The starting distribution is the number of days sickness for each employee. The shape of this distribution is not known, but it is likely to be a reverse-J (see figure 5.3(a)). Moving to the distribution of the average number of days sickness per employee of samples of 100 employees (see figure 5.3(b)), the central limit theorem states that the new distribution will be approximately normal. The parameters will be:

$$\text{Mean} = 12 \text{ days}$$
$$\text{Standard deviation} = 5/\sqrt{100}$$
$$= 0.5 \text{ days}$$

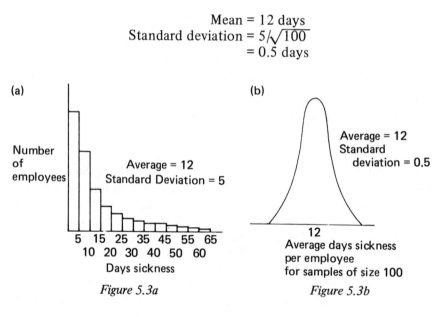

Figure 5.3a   Figure 5.3b

Estimation

The sickness record example of figure 5.3 can demonstrate how the theory can be used for estimation. Recall that estimation is the prediction of the values of population parameters given knowledge of a sample. Since the sampling distribution of the mean for samples of size 100 is normal with mean 12 and standard deviation 0.5, 95% of all such samples will have their mean in the range 11–13 days. This follows from the property of normal distributions that 95% of values lie within ± 2 standard deviations of the mean. In other words, 95% of the time (or, at the 95% confidence level) the sample mean will be within 1 day of the population mean.

This shows how a range for sample means can be estimated from the population mean. The process can be turned round to estimate the population mean from a single sample mean. Suppose, as is usually the case, the population mean was not known. However, a sample of 100 employees' sickness records gave an average number of days sickness per employee of 11.5. This must be less than 1 day away from the population mean, from the above calculations. The population must then be within 1 day of 11.5. It can therefore be said that, at the 95% confidence level, the population mean is in the range 10.5–12.5 days. Note that although the distribution of sample means is being used, in practice only one sample is taken. The mean of the sample (11.5 days) is the *point estimate* of the population mean; the range 10.5–12.5 days is the *95% confidence limits* for the estimate.

When estimating the population mean the population standard deviation was used. This is unlikely to be known. It has to be calculated from the sample, i.e. the standard deviation of the sample is calculated and used as if it were the population standard deviation. This approximation is valid provided the sample size is, as a rule of thumb, greater than 30. There are now two reasons for choosing a sample size of 30+. First, the central limit theorem will apply; second, the sample standard deviation can be used as the population standard deviation. The standard deviation of a sample mean is sometimes referred to as the standard error.

The general procedure for estimating the mean of a population is (for convenience and brevity, let the variable in question be labelled x):

- Take a random sample of size at least 30. Let the sample size be labelled n. The minimum of 30 is so that the central limit theorem holds and the standard deviation approximation is valid. A smaller sample can be used if the distribution is normal and the population standard deviation is already known. Even if these two conditions do not hold, a sample smaller than 30 can still be used, but some more advanced theory is needed.
- Calculate the sample mean ($\bar{x}$) and the sample standard deviation (s).
- The standard deviation of the sampling distribution of the mean

is calculated as

$$s/\sqrt{n}$$

- The *point estimate* of the population mean is $\bar{x}$.
- The *95% confidence limits* for the population mean are

$$x \pm 2s/\sqrt{n}$$

EXAMPLE

A sample of 49 of a certain producer's light bulbs lasted on average 1100 hours with a standard deviation of 70 before failing. Estimate the average life length for all the producers bulbs. Follow the general procedure as set out above.

- The sample, which is assumed to be random, has been taken. The size is sufficient for the central limit theorem to apply and the standard deviation approximation to be valid.
- The mean is 1100 hours, the standard deviation 70 hours.
- The standard deviation of the sampling distribution is $70/\sqrt{49}$. = 10.
- The point estimate of life length is 1100 hours.
- The 95% confidence limits are 1080–1120 hours. From normal distribution tables the limits for other confidence levels can be found. For example, since 90% of a normal distribution lies within ± 1.645 standard deviations of the mean, the 90% confidence limits in this case are

$$1100 \pm 1.645 \times 10$$

$$= 1083.55 - 1116.45$$

Note that these confidence limits were found without needing to know anything about the shape of the distribution. The only calculations were to find the mean and standard deviation from the sample.

The ideas of estimation can also help in deciding what the sample size should be in order to provide a required level of accuracy.

EXAMPLE

In the case of the light bulbs above, what sample size is needed to estimate the average length of life to within ± 5 hours (with 95% confidence)?

Instead of the sample size being known and the confidence limits unknown, the situation is reversed. At the 95% level:

$$\text{Confidence limits} = x \pm 2s/\sqrt{n}$$
$$1095 \rightarrow 1105 = 1100 \pm 140/\sqrt{n}$$
$$5 = 140/\sqrt{n}$$
$$\sqrt{n} = 28$$
$$n = 784$$

The formula used in this calculation related the accuracy of the estimate to the square root of the sample size. This is why accuracy becomes progressively more expensive. For example, a doubling in accuracy requires a quadrupling of sample size (and, presumably, expense). The trade-off between accuracy and sample size is therefore an important one.

Significance tests

Significance tests is the methodology by which it can be judged whether a particular piece of evidence is consistent with a hypothesis (a supposition about a population). The methodology involves five steps.

(1) Formulate the hypothesis. It may be any idea or hunch, the truth of which is to be investigated.

(2) Collect a sample of evidence concerned with the validity of the hypothesis.

(3) Decide on a *significance level*. This is a probability (usually and conventionally 5%) which marks the borderline between the credible and the incredible. It is supposed that if an event occurs which has a probability greater than the significance level, then it is entirely believable that it happened purely by chance. If an event occurs with a probability of less than the significance level, it is deemed too rare to have happened purely by chance. Factors other than pure chance must be at work.

(4) Calculate the probability of the sample evidence occurring under the assumption that the hypothesis is true.

(5) Compare this probability with the significance level. If it is higher, it is judged consistent with the hypothesis (it could have happened purely by chance) and the hypothesis is accepted; if it is lower it is judged inconsistent with the hypothesis (it could not have happened purely by chance) and the hypothesis is rejected.

In any circumstances (legal, medical, business, etc.) it is rare for evidence to prove conclusively the truth of a proposition. The evidence merely alters the balance of probabilities. Significance tests give a critical point which divides evidence supporting the proposition from that which does not. The significance level is this critical point. Its use is an abrupt, black-and-white method of separation, but it does provide a convention and a framework for weighing evidence.

EXAMPLE

One of the products of a dairy company is 500 g packs of butter. There is some concern that the production machine may be manufacturing slightly overweight packs. A random sample of 100 packs is weighed. The average weight is 500.4 g and the standard deviation 1.5 g. Is this consistent with the machine being correctly set and producing packs with an overall average weight of 500 g?
    Follow the steps of significance tests.

(1) The hypothesis is that the true population average weight is 500 g and there is no tendency to produce overweight packs.
(2) The evidence is the sample of 100 packs.
(3) Let the significance level be the conventional 5%.
(4) Assuming the hypothesis is true, the sample has come from a a sampling distribution of the mean as shown in figure 5.4. The sampling distribution is normal. Even if the distribution of the weights of individual packs is not normal, the central limit theorem makes the sampling distribution normal. The mean is 500 (the hypothesis). The standard deviation is the standard deviation of the individual distribution divided by the square root of the sample size.

$$\text{Standard deviation} = 1.5/\sqrt{100}$$

$$= 0.15$$

The sample taken had a mean of 500.4. Recall the way in which z values are calculated for normal distributions. In this situation:

$$z = (500.4 - 500)/0.15$$

$$= 2.67$$

From the normal table 4.3, the associated area under the

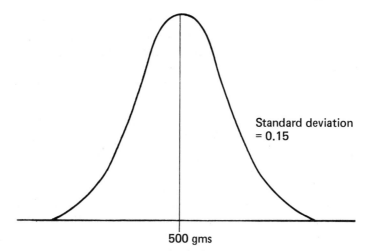

Figure 5.4

normal curve is 0.4962 (see figure 5.5). The probability of obtaining a sample result as high as or higher than 500.4 is:

$$= 0.5 - 0.4962$$

$$= 0.0038$$

$$= 0.38\%$$

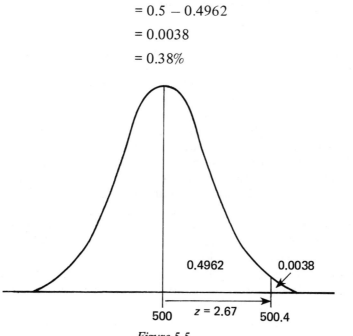

Figure 5.5

(5) The probability of the observed evidence is therefore 0.38%, much lower than the significance level. It is highly improbable that if the true average weight were 500 g, a sample with mean weight 500.4 g would be obtained. The hypothesis is rejected. The machine does appear to be producing slightly overweight packs.

This example has been based on comparing two probabilities. The probability of the sample evidence was compared with the significance level. There is an alternative procedure. The significance level can be used to calculate a *critical value*, which is a sample mean exactly on the boundary separating significant (probability less than 5%) from non-significant (probability greater than 5%) results. The acceptance or rejection of the hypothesis can then be based on comparing the sample result with the critical value. The two procedures are different ways of viewing the same thing and will always give the same result. In this example the critical value is at the point which leaves 5% in the 'tail' of the distribution. It therefore has a z-value of 1.645 (the z-value corresponding to an area in the normal table equal to 0.45). Thus,

$$\text{Critical value} = 500 + 1.645 \times 0.15$$

$$= 500.247 \text{ g}.$$

Since the sample result of 500.4 g was more extreme (further away from the mean) than this, the hypothesis is rejected. This 'critical value' method of viewing significance tests will be used in the later chapter on Correlation and Regression.

The significance test in the butter example is a so-called *one-tailed test*. Only the possibility of the machine being ill-set in the sense of producing overweight packs was considered. An inconsistent result could only be at one extreme (tail) of the distribution. However, the machine could also have been ill-set in the sense of producing underweight packs. If this possibility were also considered, the test would have to be *two-tailed*. An extreme result could arise in a sample mean being too small or too large. There is a possibility of an extreme result in either tail of the distribution. The probability calculated in step 4 should now be the probability of obtaining a sample result as far away from the mean as 0.4 g in either direction, not just 500.4 g, or higher. This new probability is double the previous one, since the area in the tail (0.0038) occurs on both sides of the mean. It is equal to 0.76%. In a two-tail test this new probability would be compared with the significance level.

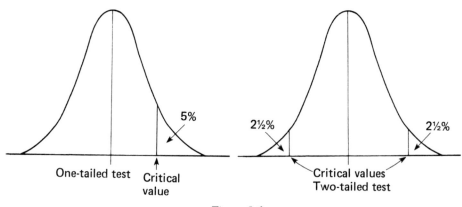

Figure 5.6

Using the critical value method, there will now be 2 critical values, one on either side of the mean. The 5% significance level is halved, 2½% in either tail. The critical values (from normal curve tables) will be approximately 2 standard deviations from the mean, at 499.7 g and 500.3 g. Figure 5.6 summarizes the difference between one- and two-tailed tests.

Worked examples

1. B. A. GROOM AND SONS

This firm of tailors have been investigating their overdue accounts. A sample of 40 overdue accounts had an average amount outstanding of £82 and a standard deviation of £20. The smallest amount overdue in the sample was £35 and the largest £480. What are the 95% confidence limits for the average amount outstanding for all their overdue accounts?

The distribution of the means of samples of size 40 is a normal distribution. In spite of the fact that the distribution of individual amounts due appears (from the range) to be skewed, the central limit theorem states that the sampling distribution of the mean will be normal. Therefore, at the 95% confidence level, the mean of the one sample collected must lie within ± 2 standard deviations of the true population mean (the average amount outstanding for all overdue accounts).

The standard deviation of this sampling distribution is equal to:

Standard deviation of individual distribution/$\sqrt{\text{sample size}}$

$= 20/\sqrt{40}$

$= 20/6.32$

$= 3.16.$

The mean of the one sample must lie within 6.32 (= 2 x 3.16) of the true population mean at the 95% confidence level. Consequently, the true population mean must be in the range:

$$82 \pm 6.3$$

$$= \text{£75.7} - \text{£88.3}$$

2. SOCIETY FOR THE UNDERSTANDING OF MATHEMATICS AND STATISTICS

The Society advises organizations how to increase the numeracy of their employees. In particular it runs short courses designed to make participants more numerate. At the end of a course a standard test is given. From its records the Society knows that the scores obtained are normally distributed with mean 520 and standard deviation 90. A new computer-based course has just been designed. A trial has been conducted with 25 participants. Their average score on the test given at the end of the course was 545. Is this consistent with the hypothesis that the computer course is equally as efficient as the old course in improving numeracy?

This is a test of the hypothesis that the new course is no different from the old and will produce an overall average test score of 520 just as before. Note that the sample size is 25, less than the usually required 30. However, since the individual distribution of scores was (and is assumed still to be) normal, the sampling distribution of the mean will be normal whatever the sample size. Also, since the standard deviation is assumed still to be 90, unchanged from before, it is not being estimated from the sample. Therefore the second reason for needing a sample size greater than 30 also does not apply.

The significance test follows the usual steps:

(1) The hypothesis is that the new course will still give an overall average test score of 520.

(2) The sample evidence is the 25 people who have undergone the computer-based course.
(3) Choose the conventional significance level of 5%.
(4) The sampling distribution of the mean is normal with mean assumed to be 520 (the hypothesis) and standard deviation equal to:

$$= 90/\sqrt{25}$$

$$= 18$$

The z-value for the sample result of an average score of 545 is thus:

$$= (545 - 520)/18$$

$$= 25/18$$

$$= 1.39$$

From the normal table 4.3, the associated area under the curve is 0.4177 (cf. figure 5.5). The hypothesis was that the new course would not change the test score. The possibility that the new course could have led to an improvement or a deterioration was recognized. The probability of the sample result must therefore be seen as the probability of a result as far from the mean as $z = 1.39$ in either direction (a two-tailed test).

Probability of sample result = 2 x 8.23% (0.0823 = 0.5 − 0.4177)

$$= 16.46\%$$

(5) This result is larger than the significance level of 5% and the hypothesis must be accepted. There is insufficient evidence to suggest that the new course makes any difference to the test scores.

Final comments

Statistical inference belongs to the realms of traditional statistical theory. Its relevance lies in its applicability to specialist management tasks, such as quality control and market research. Most managers would find that it can only occasionally be applied directly to general management problems. Its major value is that it encompasses ideas and concepts which enable problems to be viewed in broader and more structured ways.

The chapter looked at two areas: estimation and significance tests. It introduced some new theory, confidence levels and the sampling distribution of the mean which improved the usefulness and accuracy of the techniques.

The conceptual contribution that estimation makes is to push a manager to think more in terms of predicting a range rather than a single point. To take a previous market research example, the estimate that 29% of razor blades are purchased by females sounds fine. But what is the accuracy of the estimate? The 29% is no more than the most likely value. By how much could the true value be different from 29%? If it can be said with near certainty (95% confidence) that the percentage is between 27 and 31, then the estimate is a good one on which decisions may be reliably based. If the range had been 8% to 88%, then there would have been doubts about its usefulness for decision-making. Yet, surprisingly, the confidence limits of predictions are often reported with little or no emphasis.

The second area was significance tests. They are concerned with distinguishing real from apparent differences. The discrepancy between a sample mean and what is expected in the whole population is judged in the context of variation inherent in the population. An apparent difference is one that could easily have arisen purely by chance; a real difference is one that is unlikely to have arisen purely by chance and some other explanation (that the hypothesis is untrue) is supposed. Hypothesis testing draws a dividing line between the two. The dividing line marks an abrupt border. In practice, extra care is exercised over samples falling in the grey areas immediately on either side of the border.

Both estimation and significance tests can therefore improve the way a manager thinks about particular types of numerical problems. This is there their importance probably lies. As techniques, they do not generally form part of the everyday management task.

Further examples

1. LORDS INTERNATIONAL MOTOR ORGANIZATION

Lords is renowned for the high quality of the cars it produces. The company realizes that some faults do escape the strict inspection procedures, but hopes that the number is few. A recent intensive check on a random sample of 30 completed cars revealed that, on average, the cars had 10 faults. The distribution of faults appeared to be non-normal with a standard deviation, measured from the sample, of 4. What are the 90% confidence limits for the average number of defects in all cars produced?

2. PERSONAL ORDER SUPPLY TRADERS

This mail order company sends reminder letters to potential customers who have received catalogues but have not made a first order. To judge when this letter should be sent, it wishes to know accurately how long after receiving a catalogue, on average, those customers who are going to make an order do make it. A previous survey several years ago showed an average of 14 days with a standard deviation of 8 days. It is thought that both figures may have increased. What is the minimum number of customers that should be included in the survey so that the company can estimate the average number of days to make a first order to within 1 day, at the 95% confidence level?

3. T. M. IRELAND LINENS LTD.

The company makes linen and similar products. The yarns used are purchased from suppliers. One particular yarn should, according to the contract, have a tensile strength of 25 lb. However, the foreman believes that recent batches are of inferior quality. A sample of 36 specimens has a mean tensile strength of 23.75 lb and a standard deviation of 4.2 lb. Can the hypothesis that the supplier is sending yarn of mean tensile strength 25 lb be rejected at the 95% confidence level?

ANSWERS

1. Lords International Motor Organization

90% confidence limits are 8.8 to 11.2.

2. Personal Order Supply Traders

Sample size (assuming standard deviation is 8) = 256.

3. T. M. Ireland Linens Ltd.

One-tailed test at the 95% level indicates rejection of the hypothesis. The supplier is not supplying according to contract.

Relationships between Variables

Part II was concerned with the statistical analysis of single sets of numbers. In other words it was about single variables. Part III looks at situations involving more than one variable. In such situations it is the relationships between variables that are of interest. Several aspects will be discussed, including measuring the *strength* of the relationships between variables, determining their *nature*, and making *forecasts* of one variable based on its relationship with others. These three aspects are referred to as correlation, regression and business forecasting.

Preparation

Correlation, regression and forecasting are technical subjects. To attempt more than a superficial description requires some mathematics. Again, there will be some 'black box' areas. The reader with a poor or rusty mathematical background should read Appendix B (on graphs), Appendix C (on linear equations) and Appendix D (on exponents) before embarking on chapters 6 and 7. The ideas of significance tests (chapter 5) will also be used.

CHAPTER 6

Correlation and regression

By the end of this chapter the reader should know the purpose of regression and correlation and be aware of the ways in which they are used. It should be possible to criticise the output of a regression analysis and decide whether the results are good enough to use in practice. The limitations and difficulties, particularly the common-sense, non-statistical ones, which surround its use should be known. The extensions of simple linear regression are to multiple and non-linear regression.

Correlation and regression are concerned with relationships between variables. Correlation is a method for *measuring the strength* of a relationship; regression is a method for *determining a formula* expressing the relationship. Correlation shows whether a connection exists; regression finds what the connection is. For example, correlation would indicate to what extent the quarterly sales volume of a product had moved in parallel with advertising expenditure. It would show whether a high level of sales had usually corresponded to a high advertising expenditure in the same quarter (and low with low). Regression would provide the formula linking sales with advertising. The formula might look like:

Sales volume = 3000 + 3.1 x Advertising expenditure

It could then be used to predict sales in a future quarter for which the advertising level had been decided.

Initially, only correlation and simple linear regression will be described. Simple in this context means involving two variables only; linear means that the regression formula is a linear equation. Later in the chapter multiple regression, involving more than 2 variables, will be introduced.

Applications

SALES OF CHILDREN'S CLOTHING AND THE NUMBER OF BIRTHS

The two variables are firstly the annual sales of clothing for 4-year-old children at a large store, and secondly the number of births in the catchment area of the store 4 years earlier. Having records for 20 years, each set of numbers would contain 20 observations, a sales figure for each year and paired with it the number of births 4 years earlier. Such sets of numbers are called time series. A *time series* is defined as measurements of a variable at regular time intervals. Had the numbers referred not to one store over many years, but to several stores in just one year, the sets would be called *cross-sectional*. The data can be presented graphically in a *scatter diagram* (see figure 6.1). The scatter diagram suggests that there is a strong relationship between the numbers. There is a clear tendency for high sales to be associated with high births, and low with low. Correlation could measure this tendency. Regression could determine the formula relating the numbers by providing the equation of the straight line which is as near as possible to all the points. Then given the numbers of births, likely sales levels in the future could be calculated. This is a situation in which regression could be used for prediction purposes.

WEIGHT AND SALARY

The two variables are firstly the weights of a sample of executives from several organizations and secondly their corresponding salaries.

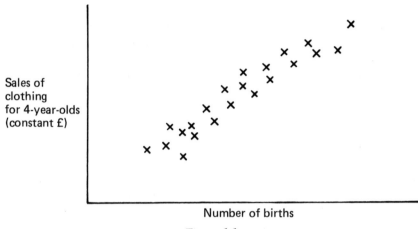

Figure 6.1

When represented by a scatter diagram the numbers appear as in figure 6.2, which shows a relationship, but not a strong one. In general, high salaries are associated with low weights, but only loosely and there are clear exceptions. There is a weak correlation between the variables. Since high salaries are associated with low weights, the correlation is *negative*. In the previous example, the correlation was *positive*, since high was associated with high. Point A is a clear exception to the general pattern and is said to be an *outlier*.

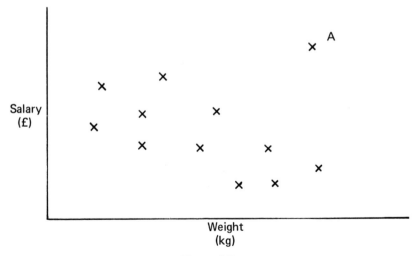

Figure 6.2

The purpose of this analysis is to gain greater understanding. Clearly, it would not be used for prediction, for example to tell an executive how many kg he should lose to secure a 20% rise in salary. The statistical results showing a correlation would be just the start of an investigation into why heavy people appear to be paid less. Although regression could be used in this case, there would be little point in doing so.

PROFITABILITY AND SIZE OF COMPANIES

The two variables are firstly the profitability (measured via the return on net assets) of a sample of 150 UK companies and the size of those companies (measured via their turnover). The scatter diagram is in figure 6.3, and reveals no clear pattern. Sometimes a high profitability is associated with large size and sometimes with low. The variables show very little correlation. This result, although seemingly disappointing, is still very useful. It shows that the line of enquiry should be varied. If the purpose here was to investigate the underlying causes of variations between companies profitabilities, then the 'low correlation' result suggests that the relationship between profitability and some other variables should be investigated.

Turnover may have some influence on profitability, but it is not a major influence. Turnover may be an inadequate measure of size and some other measure should be used; or perhaps something entirely different, such as a measure of market conditions (macroeconomic or industry variables) should be used; Or the low correlation may

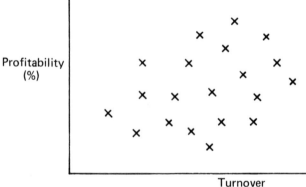

Figure 6.3

lead to the conclusion that no quantifiable explanation for variations in profitability can be found.

An intuitive approach to correlation and regression

The two questions that arise with situations of the type described above are: is there a relationship between the two variables (correlation)? If so, what is the form of the relationship (regression)? Both questions can be handled at an intuitive level. The strength of the relationship can be judged by inspection of the scatter diagram; if the scatter diagram is drawn on graph paper, a 'best' straight line can be drawn through the points and its equation measured graphically.

It should be noted that correlation and regression are concerned with approximations. Only in a very rare case (called perfect correlation) do all the points lie on a straight line (see figure 6.4). In the more usual case (figure 6.5) the points do not lie on an exact straight line and there may be several lines that seem to approximate closely to the points. Each straight line has a different equation. The three straight lines in figure 6.5 have the equations:

$$\text{Sales} = \ 3 \times \text{Births} + 4000 \quad A$$
$$\text{Sales} = \ 5 \times \text{Births} + 2000 \quad B$$
$$\text{Sales} = 10 \times \text{Births} - 2000 \quad C$$

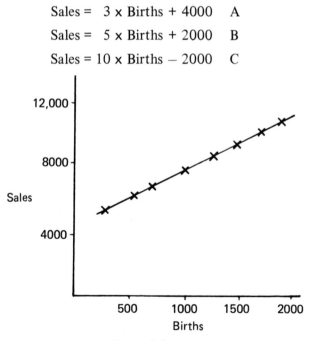

Figure 6.4

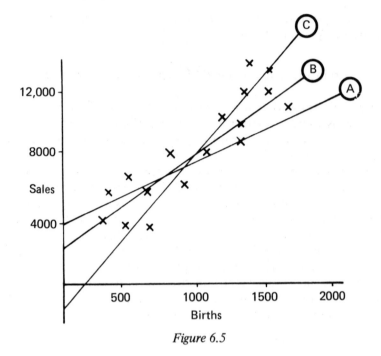

Figure 6.5

The relationship between the two variables could differ significantly depending upon which line one chose as being 'best'. Under an intuitive approach the decision as to whether the relationship is sufficiently strong and what form it takes is a subjective one. Two people could produce very different results from the same data.

Suppose one person chose line A and a second line C. If the number of births corresponding to next year's sales, births 3 years ago, are 2000 then the two persons would predict sales:

Line A: Predicted sales =   3 x 2000 + 4000 = 10,000

Line C: Predicted sales = 10 x 2000 − 2000 = 18,000

These are very different results and more subjectivity would be required to decide between them. In spite of this disadvantage the intuitive approach can be satisfactory, especially when there appears from the scatter diagram to be very little choice as to the 'best' line and when great precision is not needed (or warranted given inaccuracy in the data).

The statistical approach to the problem which is described next quantifies the strength of the relationship and provides one single 'best-fit' equation relating the two variables. It is an objective approach.

The statistical approach to linear regression

For linear regression the relationship is supposed to be a straight line. Straight lines always have an equation of the form:

$$y = a + bx.$$

Linear regression is the task of finding the values of a and b which provide the best connection between the two variables, y and x.

ALGEBRAIC AND MATHEMATICAL PRELIMINARIES

To explain the statistical approach, some mathematical symbols are needed. They may be tedious but they make for greater efficiency.

(a) *The equation of a straight line* (figure 6.6a)

Only a little mathematics is needed to show that:

a is the *intercept*, the value of y at the point at which the line crosses the y axis;
b is the *slope* of the line, the change in y for a change in x of 1 unit.

In figure 6.6a:

$$\text{slope} = b = \frac{u}{v}$$

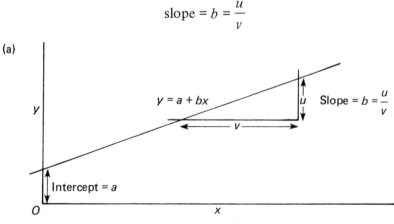

Figure 6.6a

In the case in figure 6.6a the slope is a positive number. In figure 6.6b the slope is a negative number, since in moving from A to B, *y* increases (*u* is positive) but *x* decreases (*v* is negative). Therefore:

$$\text{the ratio } \frac{u}{v} \text{ is negative.}$$

Determining the equation of a straight line amounts to finding the values of *a* and *b*. Once *a* and *b* are known, the line is completely determined. To repeat, linear regression is the task of finding the values of *a* and *b* which provide the best connection between the 2 variables.

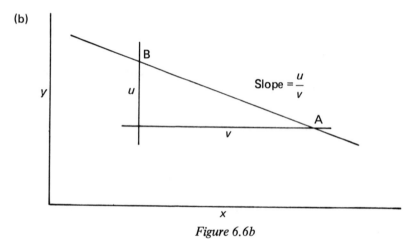

Figure 6.6b

(b) *Residuals*

Figure 6.7 is a scatter diagram. Draw *any* straight line through the set of points. In general the points will not lie on the line. Consider the first point (A). The point directly below it which does lie on the line is B. The *y* value of A is the *actual y value*; the *y* value of B is called the *fitted y value* (computed from the equation). The difference between the actual and fitted *y* values is the *residual*:

$$\text{Residual} = \text{actual } y \text{ value} - \text{fitted } y \text{ value.}$$

If the point lies above the line the residual is positive; if the point lies below the line it is negative; if the point is on the line it is zero. Each point has a residual. The residuals would be different if a different line had been drawn.

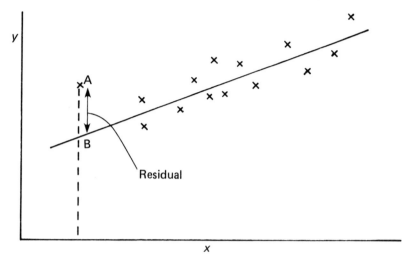

Figure 6.7

CRITERION FOR THE 'BEST-FIT'

In deciding which straight line is the 'best-fit' straight line through a set of points, a criterion is required. There must be some basis for defining in what way 'best' is best. Since the line is to be close to the actual observations the criterion should be something to do with making the residuals as small as possible. One approach would be to say that the 'best-fit' line is the one for which the sum of all residuals is a minimum compared with any other straight line drawn through these points. This does not work since positive and negative residuals will cancel with one another and so a line with large residuals can still have the sum of its residuals very small or zero. Indeed it can be proved that any line through the point defined as the mean of the x values and the mean of the y values will have its residuals summing to zero (see figure 6.8).

A second approach would be to ignore the signs of the negative residuals and make the sum of the absolute values of the residuals as small as possible. (The *absolute value* of a number means that the number is made positive even if it is negative. For example, the absolute value of +6 is +6, the absolute value of −8 is +8). This would work well except that in the past, in the absence of efficient computers, the device of taking absolute values was not easy to manipulate mathematically. For this reason this criterion has been rarely

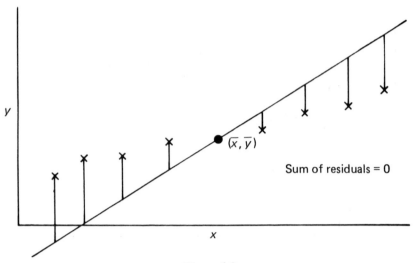

Figure 6.8

used, although the availability of computers has led to an increased usage in recent years.

The criterion traditionally employed is called *least squares*. The residuals are all squared (this eliminates the negative sign) and the sum of these squares is a minimum for the 'best-fit' line. In other words the criterion is that the 'best-fit' line is the one for which:

$$\text{Sum (residuals)}^2 \text{ is a minimum.}$$

Least squares also has some technical advantages that will not be pursued here.

FINDING THE EQUATION OF THE REGRESSION LINE

The 'best-fit' straight line for a set of points is, according to the least squares criterion, the line which has, for all possible lines, the smallest sum of squared residuals. To find the equation of this line, (to find values for a and b), a process of trial and error is *not* necessary. The 'black box' devices of differential calculus provide the following formulae:

For two variables, labelled x and y, if there are n paired observations and the regression line is of the form:

$$y = a + bx$$

then $b = \dfrac{\text{Sum } (x - \bar{x})(y - \bar{y})}{\text{Sum } (x - \bar{x})^2}$

and $a = \bar{y} - b\bar{x}$

Example

Find the regression line for the points:

x	1	2	4	5	8	$\bar{x} = 4$
y	18	17	24	23	33	$\bar{y} = 23$

Sum $(x - \bar{x})(y - \bar{y}) = (1 - 4)(18 - 23) + (2 - 4)(17 - 23)$
$$+ (4 - 4)(24 - 23) + (5 - 4)(23 - 23)$$
$$+ (8 - 4)(33 - 23)$$
$$= 15 + 12 + 0 + 0 + 40$$
$$= 67$$

Sum $(x - \bar{x})^2$ $= (1 - 4)^2 + (2 - 4)^2 + (4 - 4)^2 + (5 - 4)^2$
$$+ (8 - 4)^2$$
$$= 9 + 4 + 0 + 1 + 16$$
$$= 30$$

$$b = \frac{\text{Sum } (x - \bar{x})(y - \bar{y})}{\text{Sum } (x - \bar{x})^2}$$

$$= \frac{67}{30}$$

$$= 2.23$$

$$a = \bar{y} - b\bar{x}$$

$$= 23 - (2.23) \times 4$$

$$= 14.08$$

$$y = 14.1 + 2.23x$$

The statistical approach to correlation

The formulae for *a* and *b* can be applied to any data set. The scatter diagram may show that the points lie on a circle, but the 'best-fit' line can still be found. Quantifying the strength of the correlation will help to decide whether the closeness of the points to a straight line is sufficient to warrant finding the regression line.

The strength of the correlation is measured by the *correlation coefficient* (denoted by *r*) which is a number between −1 and +1. The meaning of the different values of the correlation coefficient is illustrated in figure 6.9.

The correlation coefficient can take on all values between −1 and +1 and the five illustrations are just examples of the situations that could occur. Correlation coefficients close to −1 or +1 always indicate a strong relationship between the variables, but its exact interpretation depends upon the number of observations. For 5 observations, $r = 0.6$ would not indicate a strong correlation; for 200 observations, $r = 0.6$ would indicate a strong correlation. The reason for this is that when two variables are not related it is easier to fit a straight line to 5 points than 200. Therefore the standard against which to judge the strength of the correlation is stricter for 5 observations than for 200. Tables in statistical textbooks show whether the *r* calculated for a given number of observations indicates a strong correlation or not. This procedure will be briefly discussed later.

The formula for calculating the correlation coefficient is:

$$r = \frac{\text{Sum } (x - \bar{x})(y - \bar{y})}{\sqrt{\text{Sum } (x - \bar{x})^2 \times \text{sum } (y - \bar{y})^2}}$$

The reason *r* does measure the strength of a relationship is a mathematical one, but a partial understanding can be gained by considering the correlation coefficient squared. By convention, and for no apparent reason, it is written as R^2.

Before carrying out a regression analysis, the total variation in the *y* variable can be measured by:

$$\text{Total variation} = \text{Sum } (y - \bar{y})^2$$

This expression measures variation in a way that is similar to the variance.

After a regression analysis, one can think of two types of variation.

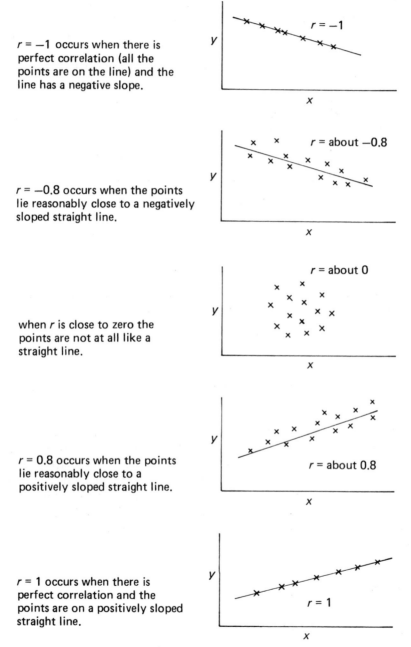

$r = -1$ occurs when there is perfect correlation (all the points are on the line) and the line has a negative slope.

$r = -0.8$ occurs when the points lie reasonably close to a negatively sloped straight line.

when r is close to zero the points are not at all like a straight line.

$r = 0.8$ occurs when the points lie reasonably close to a positively sloped straight line.

$r = 1$ occurs when there is perfect correlation and the points are on a positively sloped straight line.

Figure 6.9

Firstly, there is the 'explained' variation – the square of the differ-ence between the fitted y value and the mean y value:

$$\text{Explained variation} = \text{Sum (fitted } y \text{ value} - \bar{y})^2$$

It is called 'explained' because this variation is understood, being on account of the straight line relationship between y and x.

Secondly, there is the 'unexplained' variation which is the variation in the residuals. This part of the variation has not been explained, since it is 'left over' from the best-fit line.

$$\text{Unexplained variation} = \text{Sum (residual)}^2$$

Using some "black box" mathematics, it is possible to show that:

$$\text{Total variation} = \text{Explained variation} + \text{Unexplained variation}$$

and also that:

$$R^2 = \frac{\text{Explained variation}}{\text{Total variation}}$$

This indicates why the correlation coefficient works as a measure of the strength of a relationship. If $R^2 = 1$, then

$$\text{Explained variation} = \text{Total variation}$$

and $$\text{Unexplained variation} = 0.$$

If unexplained variation = 0, then the sum of the squared residuals is 0 and therefore all the residuals must each be 0. The points all lie on the line.

The essence of the correlation coefficient is then that on carrying out a regression analysis the total variation in the y values is split into two parts: (a) a part that is explained by virtue of associating the y values with x values, and (b) a part that is unexplained since the association is not an exact one. The correlation coefficient tells what proportion of the original variation has been 'explained' by drawing a line through the points. The higher the proportion, the stronger the correlation.

EXAMPLE

x	1	2	4	5	8
y	18	17	24	23	33

This is the same example used for regression. Some calculations have already been made.

Sum $(x - \bar{x})(y - \bar{y}) = 67$

Sum $(x - \bar{x})^2 \quad = 30$

Sum $(y - \bar{y})^2 \quad = (18 - 23)^2 + (17 - 23)^2 + (24 - 23)^2$
$+ (23 - 23)^2 + (33 - 23)^2$

$= 25 + 36 + 1 + 0 + 100$

$= 162$

$$r = \frac{\text{Sum } (x - \bar{x})(y - \bar{y})}{\sqrt{\text{Sum } (x - \bar{x})^2 \times \text{Sum } (y - \bar{y})^2}}$$

$$= \frac{67}{\sqrt{30 \times 162}}$$

$$= \frac{67}{69.71}$$

$$= \underline{0.96}$$

The correlation coefficient is 0.96, indicating a strong relationship between the variables.

Residuals

The correlation coefficient is one test for deciding whether the two variables are linked. By itself, however, this test is not sufficient to indicate a strong relationship. Figure 6.10 shows a scatter diagram

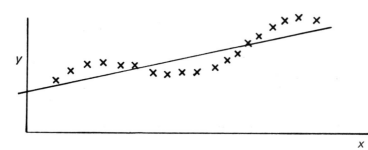

Figure 6.10

for which the correlation coefficient is high, but for which the best-fit line does not represent the way in which the variables are linked.

The situation in Figure 6.10 may occur where there is a seasonal or cyclical pattern in the data. To ensure that the straight line model does adequately represent the relationship between the variables, the residuals should be examined. The residuals should be random. Random means the residuals are in no particular order.

If the underlying relationship between the variables really is a straight line, then one would expect that the reason the points do not fall exactly on the line is that there is distortion on account of several minor effects, for example measurement errors and one-off fluctuations. Since effects like these occur randomly, the residuals should be random, and there should be no pattern in them. In figure 6.10 there is a pattern in the residuals, each having a link with previous ones. This is called *serial correlation*.

A further aspect of randomness is that the residuals should have constant variance. There should be no tendency for the residuals to be more widely scattered at some parts of the line than others. If they are as in figure 6.11 then there is a pattern in them and they are not random. If the residuals have constant variance they are said to be *homoscedastic*; if not, they are heteroscedastic.

The purpose of testing the residuals for randomness can be summarized as follows. If the residuals have a pattern to them, then this pattern should be incorporated (in a way as yet unspecified) into the model; if the residuals are random, there is no pattern to them and they are unpredictable. In the latter case the best-fit line explains,

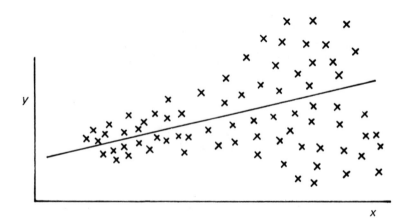

Figure 6.11   Heteroscedasticity

therefore, as much about the data as possible. What is left over, the unexplained variation, cannot be explained because the residuals are random.

The first way of testing for randomness is visual. The scatter diagram, with the best-fit line drawn in, is inspected for patterns such as those in figure 6.10 and 6.11. There are also statistical tests for randomness.

The test on the residuals for randomness is often omitted from regression analysis and only the correlation coefficient used as a test of goodness of fit. This leads to an incomplete understanding of the relationship and mistakes in its use and interpretation.

Causality

What has been dealt with so far has been the statistical aspects of regression analysis. Potentially more important are the interpretative aspects. Even when a relationship between variables satisfies all the statistical tests, what can be concluded and how should the results be used?

The essential fact to remember is that while the statistical results show whether the variables are associated, they tell very little about causality. Extreme examples are often used to illustrate this point.

For example, there is a close association (high correlation, random residuals, etc.) between the price of rum and the level of remuneration of the clergy. This does not mean that a rise in salary for the clergy will be spent on rum, thereby depleting stocks and *causing* a rise in the price. What it probably means is that there is a third factor (inflation? or a general increase in society's affluence?) which affects both variables so that in the past the salaries and the price of rum have moved together. It would be a mistake to suppose that if conditions were to change the relationship must continue to hold. For example if the clergy, for some good philanthropic purpose, agreed next year to take a salary cut, then the situation is substantially different from anything that has gone before. The price of rum would continue to change in response to inflation or whatever the third factor was, and the association would be broken. Had the relationship been a causal one, then manipulating one variable would cause a corresponding change in the second variable. It is difficult to be sure that a relationship is causal as well as associative. One approach is a negative one. A third factor (as in the above example), the existence of which suggests a relationship is *not* causal, is sought.

The earlier example of the link between the birth rate in an area and the sales of 4-year-olds' clothing 4 years later is more likely to be a causal one (although there will be more influences on sales than the number of children needing clothes). A new situation, such as a decrease in the number of births, could be expected to produce a drop in sales 4 years later.

Mostly the existence of causality over and above statistical association is more difficult to determine than in the above two relatively clear-cut examples. The most notorious case is the alleged link between lung cancer and smoking. There is certainly a strong statistical association between the two, but causality is a different matter. The tobacco companies say that there is a third factor, perhaps the level of stress in society, which makes people smoke more while at the same time causing greater incidence of lung cancer. If this were true, then telling people to stop smoking would not only not affect their likelihood of suffering from lung cancer, but would make them less able to cope with life's stresses (if indeed smoking does help in this regard).

Usually the question of causality can be advanced by collecting new data from new situations and showing that the relationship still holds. For instance, data on women smoking show a link with lung cancer even though the numbers of women smoking, unlike men, were very low until after the Second World War. New data like these do not furnish proof, but strengthen existing evidence.

Regression and correlation are probably at their most useful when testing relationships whose causality is suggested by some theory. For example, if a medical mechanism by which smoking could cause lung cancer were suspected, then regression and correlation would supply strong supporting evidence.

Sampling nature of regression and correlation

How to calculate the 'best-fit' line through a set of points and test the strength of the relationship was described in a non-rigorous fashion. A more rigorous context for the calculations would be as follows.

The statistician believes that in the whole population of observations a straight-line relationship does exist between the two variables and that any deviations from this (the residuals) are caused by many minor random disturbances. He then takes a sample of observations of the two variables in order to estimate the coefficients (the values of *a* and *b*) of the equation. The sample is merely the set of points upon which the calculations of *a* and *b* were based earlier in this chapter. To verify his belief in a straight-line relationship the statistician has to measure the correlation coefficient (*r*) and check the residuals. Note that *r*, *a* and *b* are all referred to as coefficients, of one sort or another.

However, the fact that these calculations are made from a sample, means that the values of *r*, *a* and *b* are estimates. If a different sample had been chosen, different values would be obtained. If many samples were taken distributions of *r*, *a* and *b* could be formed; *r* (and *a* and *b*) has not a single fixed value, but is a variable with a distribution of its own. In practice only one sample is taken and this is used to estimate the distributions. The variability of the distributions is estimated from the variability of the residuals in the chosen sample.

The distribution is the basis of the test to determine whether *r* is sufficiently high to indicate a close relationship. Consult the bibliography for details of this test.

Regression on a computer

If correlation and regression relating a company's sales volume and advertising expenditure were carried out on a computer, the results might be displayed as in table 6.1. A linear relationship between sales volume and advertising expenditure would be of the form:

$$\text{Sales volume} = a + b \times \text{Advertising expenditure}$$

The column headed 'Coefficient' gives the *a* and *b* values for this equation. The relationship is:

$$\text{Sales volume} = -1.0 + 3.7 \times \text{Advertising expenditure}$$

The correlation coefficient (*r*) is printed as 0.89. Because of the

Table 6.1

Variable	Coefficient	Standard error
Advertising	3.7	0.6
Constant	−1.0	0.4

Correlation coefficient = 0.89

sampling nature of correlation and regression, a and b have distributions. If a different selection of points had been chosen, different values of a and b would have been calculated. Since a and b have a distribution, they have standard deviations (in this context often referred to as standard errors), which are displayed in the column headed 'Standard error'. Use will be made of these standard errors later.

Multiple regression analysis

The regression investigated so far has been simple linear regression analysis. A dependent variable (y) has been related to an independent variable (x), by an equation of the form:

$$y = a + bx.$$

The basic idea can be extended so that there is more than one independent variable and the relationship takes a form such as, for example:

$$y = a + b_1 x_1 + b_2 x_2 + b_3 x_3$$

Here there are 3 independent variables x_1, x_2, x_3. Multiple regression analysis is the means by which the values of the constant a and the three coefficients, b_1, b_2 and b_3 are estimated. The criterion used to make the estimates is the same, minimizing the sum of squared residuals, but inevitably the formulae are more complicated. A computer would always be used to make the calculations.

EXAMPLE

An early example was of predicting sales of 4-year-old children's clothing at a retail store from the number of births in the store's catchment area 4 years earlier. A further factor that may affect the sales is the level of prosperity in the community. One way of measuring this would be an economic variable such as the gross domestic product (GDP). Sales might then be predicted by relating them to the number of births four years earlier *and* the GDP.

$$\text{Sales} = a + b_1 \times \text{Births} + b_2 \times \text{GDP}$$

If there were, say, a 30-year record of each variable and it were fed into a computer package, the results might look like Table 6.2. From

Table 6.2

Variable	Coefficient	Standard error
Births	2.12	0.41
GDP	0.73	0.22
Constant	7.92	1.91
Correlation coefficient = 0.96		

this table the equation relating the variables can be written (in un-specified units) as:

$$\text{Sales} = 7.92 + 2.12 \times \text{Births} + 0.73 \times \text{GDP}$$

The correlation coefficient is used as in simple regression. Just as before its square (R^2) measures the proportion of variation explained.

Just as in the case of simple regression, the residuals should be inspected for randomness. Even though the relationship cannot be graphed because there are no longer two variables, the residuals are still calculated as the difference between actual and fitted y values.

DECIDING WHICH VARIABLES TO INCLUDE

However, multiple regression does require some extra tasks. In multiple regression, especially with many independent variables, the question arises: should all the independent variables be included in the equation? The question can be answered by carrying out a series of significance tests. For each independent variable it is hypothesized that the true value of its coefficient is zero. The variable is supposed to have no effect on the dependent variable. The test is based on the standard error (known from the computer output) and on making the assumption that the distribution of the coefficient is normal. There are good reasons, not specified here, for being able to make this assumption. A significance test is carried out for each variable. If the hypothesis (that the coefficient for the variable is zero) is rejected then the variable has been rightfully included in the formulation since it does have some effect. If the hypothesis is accepted then the variable may be omitted from the equation, since it has no effect. Sometimes the variable is still included in the equation, usually because there are non-statistical reasons for continuing to include it and because the test results were close to the accept/reject borderline.

In the example relating sales to births and GDP the significance tests for each variable would both lead to rejection of the zero coefficient hypothesis. Since ± 2 standard errors covers 95% of a normal distribution, in a 5% significance test the critical values for births are ± 0.82 (= 2 standard errors). The critical values for GDP are ± 0.44 (= 2 standard errors). The coefficient estimates for the variables are 2.12 and 0.73 respectively, therefore the hypothesis is rejected in both cases and both variables are rightfully included in the formulation.

A further check that has to be made in multiple regression is to guard against the problem of *multi-collinearity*. This occurs when two or more of the independent variables are highly correlated with one another. When this happens the two variables do not bring separate sets of information to the equation. They bring, essentially, the same information. The consequence is that the coefficients of the variables are unreliable since the regression procedure finds it difficult to discriminate between their influence on the dependent variable. Consequently they cannot be used to judge the magnitude of the effect of either of the variables on the *y* variable. Small changes in the number of observations can bring about large changes in the values of the estimated coefficients.

The test for multi-collinearity is to look at the correlation coefficients between all of the variables taken in pairs. The remedy is (a) to use only one of the variables, (b) to amalgamate the two variables in some way or (c) to substitute one of the variables with a new variable which is independent of the other.

As an example, consider a situation where there are four independent variables. The correlations between them (the correlation matrix) are shown in Table 6.3. The correlation coefficients are all small, except for on the main diagonal (which *have* to be 1.0) and the correlation between variables 2 and 3 which is 0.8. Variables 2 and 3 are highly correlated because they contribute two very similar sets of

Table 6.3

Variable	1	2	3	4
1	1.0	0.1	0.2	0.2
2		1.0	**0.8**	0.3
3			1.0	0.1
4				1.0

information. These variables are collinear. The three remedies to multi-collinearity mentioned above should be investigated to see which is appropriate in the case. Again this example merely touches on a wide area. The bibliography references should be consulted for more detail.

Table 6.4 summarizes the similarities and differences between simple and multiple regression.

Table 6.4

Similarities	Differences
Use of R^2 to measure the strength of the relationship	Significance tests to determine which variables should be included
Inspection of residuals for randomness.	Check for multi-collinearity

Non-linear regression

One extension of simple linear regression is in the direction of multiple regression. Another area of extension is to non-linear regression in which the variables may appear in squared, logarithmic or other terms and the scatter diagram is no longer a straight line. There are two main ways in which 'curved' relationships can be handled.

The first is *curvilinear* regression. This is the case when a curved relationship is handled as if it were linear multiple regression. For example the equation:

$$y = a + bx + cx^2$$

is a curved relationship, between y and x, with a, b and c as constants, as in figure 6.12. Curvilinear regression treats this equation as if it were of the form:

$$y = a + bx + cz$$

In other words x^2 is treated as if it were an entirely separate variable, rather than the square of x. Multiple regression is used to estimate a, b and c as if the equation were linear. x^2 is restored to the equation which can then be used for prediction.

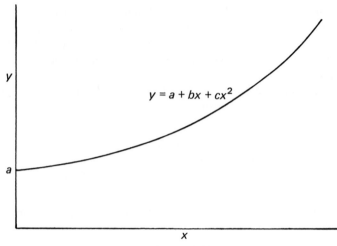

Figure 6.12

The second extension to curved relationships is by use of *trans-formations*. For example a relationship of the form:

$$y = ae^{bx} \quad (e \text{ is the constant} = 2.718 \ldots)$$

is not linear between y and x. This is an *exponential function*. It is characterized by the fact each time x increases by 1 unit, y increases by a constant proportion of itself. Compare this with a linear function ($y = a + bx$) where an increase of 1 unit in x results in y increasing by a constant amount ($= b$). There are clearly situations in which an exponential function might apply, such as the increase in the sales of some new product over time. If logarithms are taken of each side, the equation becomes:

$$\log y = \log (ae^{bx})$$

$$\log y = \log a + bx$$

using the rules for manipulating logarithmic functions. The new equation is linear between $\log y$ and x with coefficients $\log a$ and b. Figures 6.13 and 6.14 show the effect of the transformation, after which the methods of linear regression are used to estimate $\log a$ and b. From $\log a$, the value of a can be found and therefore the exact form of the relationship between y and x.

There is a wide range of possible transformations of which the above logarithmic transformation is just one.

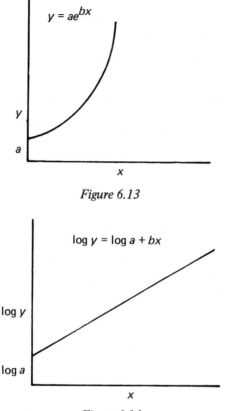

Figure 6.13

Figure 6.14

Some reservations about regression and correlation

(1) The most frequent cause of error is possibly the confusion between statistical association and causality, previously discussed.

(2) The statistical complexity of regression analysis makes it easy to make statistical judgements to the exclusion of common-sense ones. This can lead to the acceptance of *spurious* correlations. A spurious correlation is one where the correlation coefficient is high but where there is in fact no true underlying correlation. This may arise purely by chance when a sample taken at random from uncorrelated variables 'just happens' to have a high correlation. In other words, although selected at random, the sample is unrepresentative.

Spurious correlations may also occur because of a fault in the model underlying the regression analysis. For instance, the regression may be based on the equation:

$$\text{Return on net assets} = a + b \times \text{Profit}$$

with observations taken over a sample of companies. Such a regression has an inbuilt likelihood of a high correlation coefficient. Since:

$$\text{Return on net assets} = \frac{\text{Profit}}{\text{Net assets}}$$

the regression equation is:

$$\frac{\text{Profit}}{\text{Net assets}} = a + b \times \text{Profit}$$

Profit appears in both the y variable and the x variable. There is a strong likelihood therefore (especially if there are only small variations in net assets from company to company) that high y values will automatically correspond to high x values and low with low. A high correlation coefficient will result, but it would not be possible to draw any serious conclusions from this.

(3) *Extrapolation* should be avoided if possible. Extrapolation means using the equation for prediction outside the range of conditions or data on which the coefficients were estimated. For example, if a regression model were based on x values in the range 50 → 100, then forecasting the y value for $x = 200$ is extrapolation. It is dangerous since the model is based on the effects on y when x varies between 50 and 100. Nothing is known as to what happens to y when x is in the region of 200.

(4) Regression applies to only single sets of data. The data in figure 6.15a may have a high correlation coefficient hiding the fact that

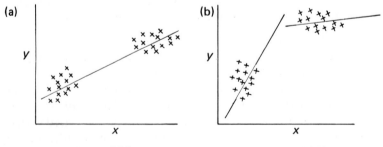

Figure 6.15a   Figure 6.15b

two separate straight lines (figure 6.15b) would be more appropriate. This highlights the need to look carefully at data, especially scatter diagrams.

(5) The least-squares criterion can be misleading by being over precise. The three lines A, B, and C in figure 6.16 may be very close to one another in terms of variation explained (perhaps less than 1% difference), although their equations are very different. The least-squares principle will pick out one of the equations as best, giving the impression that it is clearly the best when other equations are to all practical purposes as good, and may be better, when non-statistical factors are considered. Prior knowledge or previous work may be more relevant in deciding between A, B and C but such possibilities are not included in the statistical theory.

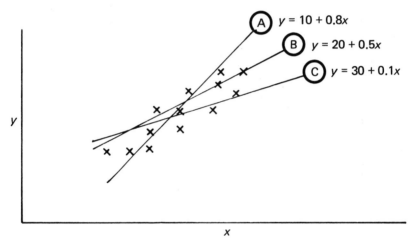

Figure 6.16

(6) It is difficult to generalize and extend a regression equation to a different set of data. A regression line always passes through the mean of a set of data. If new data are collected the regression line will only be the same if the two sets of data have the same means, which is very unlikely. Therefore, a different line is obtained when new data are introduced. The theory of regression does not show how to deal with situations like this. Should the new line be used, or the original, or a combination? All three are likely to give very different forecasts. Tests judge whether one line is significantly (in the

statistical sense) close to the other, but the more important question of which line to use is another matter.

(7) Interchanging the x and y values gives a different regression line. There is a difference between the regression of y on x (minimizing the sum of squared y residuals) and the regression of x on y (minimizing the sum of squared x residuals) – (see figure 6.17). It is not always clear which way round is correct. A judgement has to be made as to which variable is more important (and should be the one whose residuals are minimized).

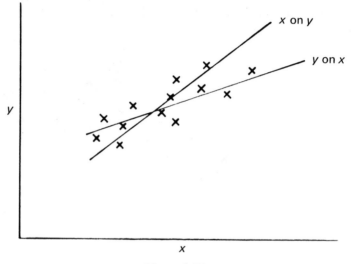

Figure 6.17

Worked examples

1. H. A. MacDUFF SERVICES

A pilot study involving 5 theatre agency booking offices provides the following information about the number of transactions (y) in hundreds and the number of clerks (x).

y:	5	6	7	10	12	Mean = 8
x:	2	1	6	9	7	Mean = 5

Draw a scatter diagram (Fig. 6.18). Does the relationship look linear? Calculate the correlation coefficient. Calculate the regression line of y on x, the residuals from this line, and their variance.

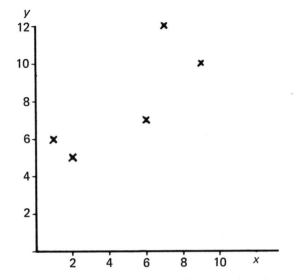

Figure 6.18   Scalter diagrams – H. A. MacDuff Services (see Example 1)

High y values tend to go with high x values. The underlying relationship could be linear with a lot of scatter.

$$\text{Sum } (x - \bar{x})^2 = (2-5)^2 + (1-5)^2 + (6-5)^2 + (9-5)^2 + (7-5)^2$$
$$= 9 + 16 + 1 + 16 + 4$$
$$= 46$$

$$\text{Sum } (y - \bar{y})^2 = (5-8)^2 + (6-8)^2 + (7-8)^2 + (10-8)^2 + (12-8)^2$$
$$= 9 + 4 + 1 + 4 + 16$$
$$= 34$$

$$\text{Sum } (x - \bar{x})(y - \bar{y}) = (5-8)(2-5) + (6-8)(1-5) + (7-8)(6-5)$$
$$+ (10-8)(9-5) + (12-8)(7-5)$$
$$= 9 + 8 - 1 + 8 + 8$$
$$= 32$$

Correlation coefficient $= \dfrac{\text{Sum } (x - \bar{x})(y - \bar{y})}{\sqrt{\text{Sum } (x - \bar{x})^2 \times \text{Sum } (y - \bar{y})^2}}$

$$= \dfrac{32}{\sqrt{46 \times 34}}$$

$$= 0.81$$

Slope coefficient $= \dfrac{\text{Sum } (x - \bar{x})(y - \bar{y})}{\text{Sum } (x - \bar{x})^2}$

$$= \dfrac{32}{46}$$

$$= 0.70$$

Intercept $= \bar{y} - b\bar{x}$

$$= 8 - 5 \times 0.7$$

$$= 4.5$$

The line is $\underline{y = 0.7x + 4.5}$

Points		Line	
		Fitted	
x	*y*	*y*	*Dev.*
2	5	5.9	−0.9
1	6	5.2	+0.8
6	7	8.7	−1.7
9	10	10.8	−0.8
7	12	9.4	+2.6

Variance $= \dfrac{(-0.9)^2 + (0.8)^2 + (-1.7)^2 + (-0.8)^2 + (2.6)^2}{4}$

$$= \dfrac{11.7}{4}$$

$$= \underline{2.9}$$

The calculated variance of 2.9 is the lowest for any straight line because of the least-squares criterion. In practice, one would need more observations than 5 to do regression analysis. This example is intended to illustrate calculations. In a more practical example it would be necessary to judge how 'sensible' the model is and whether the residuals have the requisite properties.

2. R. A. DISH

The average level of sales per week (Y) for a sample of 42 Dish retail food establishments and a measure of disposable income per family in each establishment's catchment area are given in table 6.5. The stores have been numbered for convenience. The computer gives the following results after regressing Sales Y on family income X:

Variable	Coefficient	Standard error
X	0.173	0.019
Constant	37.140	5.971

Correlation coefficient = 0.78

Table 6.5

Store number	Average weekly sales (£000s) Y	Average disposable family income (coded) X
1	90	301
2	87	267
3	86	297
4	84	227
5	82	273
.	.	.
.	.	.
.	.	.
39	64	157
40	61	141
41	58	119
42	52	133
Mean	74.7	217

A listing and a scatter diagram of the residuals are as shown in figure 6.19. Note that the scatter diagram to show the residuals is not *y* against *x*, but residuals against fitted *y* values. This is the usual practice and allows for the multiple regression case where it is not possible to plot *y* against several *x* variables. Store numbers are circled on the scatter diagram.

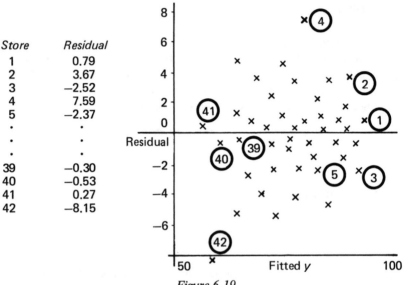

Store	Residual
1	0.79
2	3.67
3	−2.52
4	7.59
5	−2.37
.	.
.	.
.	.
39	−0.30
40	−0.53
41	0.27
42	−8.15

Figure 6.19

(1) What is the estimated relationship between sales and income?
(2) How good is the fit of the relationship?
(3) Why does the regression coefficient *b* have a standard error?
(4) What would you predict for sales for a store in an area where the disposable income per family is 221?
(5) What reservations would you have about forecasting sales in this way?

(1) The estimated relationship is:

Sales = 37.14 + 0.173 x Disposable income

(2) The fit is good since the correlation coefficient (*r*) is 0.78. R^2 is 0.61, therefore 61% of the original variation in sales has been explained.

To further examine the goodness of fit the residuals should be inspected for randomness. There is no obvious pattern either in the scatter diagram or in the listing of residuals, which appear to be random. Statistical rests for randomness (not dealt with here) could provide confirmation of this. Nor is there any sign of heteroscedasticity, since the variance of the residuals appears to be approximately the same at all fitted *y* values.

(3) Regression analysis is based on sampling. If a different set of observations (different stores or different time periods) had been used different results would have been obtained. The slope coefficient can vary from sample to sample. It therefore has a standard error which is estimated from variation within the particular sample chosen.

(4) The sales forecast for *x* = 221

$$Sales = 37.14 + 0.173 \times 221$$

$$= 75.373$$

i.e.   £75,373

(5) The relationship passes the statistical tests. Non-statistically, it is perfectly possible that families' average income in a catchment area does have a bearing on the sales of a store. In other words, the equation makes sense. On the other hand, there are many more influences on a stores sales than family income, and these are not included in the forecasting method.

The quality of the data may give rise to doubts. While store sales should be relatively easy to measure accurately, average family income will be difficult to measure. It would be important to ensure the information was up to date. The catchment area will be difficult to define and this could make the data inaccurate.

3. P. H. DEE'S REPORT

Research work in the area of health care included a correlation between the number of times as adults a random sample of 731 patients had visited their doctors, and the age of those patients at death. The resulting correlation coefficient was high. What are your comments on the researcher's conclusion that visiting a doctor prolongs life?

The high correlation coefficient means that the two variables have a high level of statistical association. It does not demonstrate a causal relationship. The researcher, by concluding that visiting a doctor prolongs life, is assuming a causal relationship along the lines:

More visits to doctors $\rightarrow$ Longer life

If there is a causal aspect to the relationship it could equally well be argued that the relationship works:

Longer life $\rightarrow$ More visits to doctors

since the longer a person lives the more opportunity he or she has to make visits to doctors. The researcher is making a big jump from associative evidence to a causal conclusion without justifying it.

Final comments

Regression and correlation are important techniques with a wide range of applications in macroeconomics (forecasting employment, exchange rates, GDP, etc.), finance (setting budgets), corporate planning (sales forecasting, manpower planning) and other areas. The statistical theory is extensive and this sometimes leads to it wrongly dominating the analysis. Many major errors in using regression and correlation seem to occur because common sense is not applied, as in the case of the third worked example.

To summarize, the stages in a regression analysis are:

(1) Check that the variables included in the regression equation are the right ones. This means:

- The equation is conceptually correct. The independent variables included should be the ones which, *a priori*, one would expect to affect the dependent variables. Situations such as clergy salaries *vs*. price of rum should be avoided.
- The variables should not lead to spurious correlations.
- The variables should not be collinear.
- The data should be inspected using scatter diagrams to determine whether the regression should be linear or non-linear.

(2) Estimate the coefficients (a, b, etc.) of the model, and produce other statistics such as correlation coefficient, standard errors, etc.

(3) Check that the correlation coefficient is high.

(4) Check that the residuals do not have the characteristics of:

- serial correlation;
- heteroscedasticity.

(5) In multiple regression check that there is no multi-collinearity

and use significance tests to confirm that the correct variables have been included.

The model should now be ready to use, but of course each situation is different and the above should be regarded as a basic, not a comprehensive, checklist.

This chapter has outlined the principles of regression and correlation. A manager should understand how these principles are applied without himself needing to become a practitioner. The contribution a manager can make to a regression study is his knowledge of non-quantifiable factors. However, he will only be able to make this contribution if he knows sufficient of the concepts (not details) to have the confidence to participate in discussions with the practitioners. When the management element is brought into statistical studies, the chances of a successful outcome will increase enormously.

Further examples

1. LONDON AREA TRANSPORT EXECUTIVE

The Executive wishes to determine whether there is any relationship between the age of a bus and its annual maintenance costs. A sample of 15 buses produced the data shown in table 6.6. What is the correlation coefficient?

Table 6.6

Age of bus (years)	Annual maintenance cost (£)
1	550
1	600
2	650
2	600
2	750
2	850
3	800
3	750
3	900
4	1050
4	850
4	950
4	1300
5	950
5	1200

2. R. A. DISH (continued)

In the earlier example with this company, an attempt was made to try to explain sales (Y) in terms of disposable income (X_1). Another possible independent variable was later added to the study. This variable was the number of families in each establishment's catchment area. A computer package regressed sales (Y) against average disposable income (X_1) *and* number of families (X_2). The results are given below.

Variable	Regression coefficients	Standard error
X_1	0.087	0.028
X_2	0.058	0.019
Constant	26.240	4.792
Correlation coefficient 0.96		

Correlations between variables:

	Y	X_1	X_2
Y	1	0.91	0.89
X_1		1	0.77
X_2			1

From the above information is the multiple regression model better than the simple regression model presented earlier for predicting sales? What further checks should be made?

3. ROYAL UNITED SCRAP TRADERS

The company has 12 plants at which scrap metal is processed. The plants have different capacities and, perhaps because of this, the average processing cost per ton varies between plants (see figure 6.20).

The data came from recent accounting information shown in table 6.7. The company wished to estimate the average processing costs for

Table 6.7

Plant	Capacity (tons/week)	Average cost/ton (£)
1	200	61.2
2	100	92.0
3	250	51.1
4	350	53.2
5	280	54.8
6	275	52.2
7	400	47.4
8	150	72.1
9	200	51.8
10	250	49.6
11	240	48.2
12	300	46.1

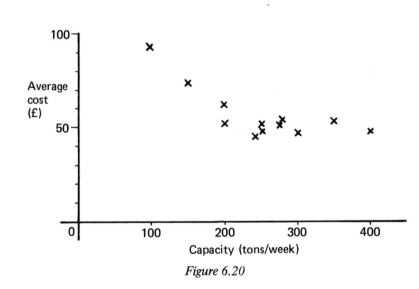

Figure 6.20

new plants they intended to build by relating the average cost to the capacity. A scatter diagram of the data is shown in figure 6.20. The computer results from regressing average cost/ton against capacity are shown below:

Variable	Regression coefficient	Standard error
1	−0.124	0.032
Constant	87.511	13.301
Correlation coefficient −0.77		

Should the company use the regression equation to predict costs at new plants? What could be done to improve the model?

ANSWERS

1. London Area Transport Executive

$r = 0.83$.

2. R. A. Dish

The multiple regression model is better because:

- it has a higher r value;
- both X_1 and X_2 have coefficients which are significantly different from zero.

Note that there is a danger of collinearity since X_1 and X_2 have a high correlation ($r = 0.77$). These coefficients may not be reliable. In addition, the residuals should be inspected for randomness.

3. Royal United Scrap Traders

The regression should not be used for the following reasons:

- The scatter diagram suggests the relationship is curved rather than a straight line.
- There is a danger of a spurious regression since average costs (= costs/tons) is being regressed against capacity. If plants are all being operated at about the same percentage of capacity the equation has an in-built tendency to a negative correlation.

The company should eliminate the spurious element by regressing total costs against capacity (assuming that all plant tonnage is approximately the same percentage of its capacity). If the scatter diagram for costs *vs.* capacity shows a curved effect, then a transformation should be used. Other variables, such as the age of the plant, might also be introduced as additional independent variables.

Business forecasting

By the end of the chapter the reader should be able to do some practical forecasting using any of several techniques. He should be aware of the substantial management aspects which are so important in forecasting. Testing and comparing different techniques is also described.

In recent years forecasting seems to have become more and more difficult. Data, whether from companies, industries or countries, seem to be increasingly volatile. The increasing uncertainty is evident in the business world and interest in making use of forecasting methods is growing. The rewards for good forecasting are very high. At the same time, the cost of bad or non-existent forecasting is greater than ever. Indeed some of the most expensive management errors have been made as a result of the poor use of forecasting methods. Some of the errors occurred because of technical mistakes, but many were made because of the way forecasting was organized and managed.

    The purpose of this chapter in two-fold: firstly, to discuss the management of forecasting methods within an organization; secondly, to review forecasting techniques and ways of testing their accuracy. The review is not limited to quantitative methods of forecasting and will include some qualitative techniques.

    Most people know how forecasting can be used. Instead of starting with some applications of forecasting, some forecasting errors will be described. They should be regarded in the positive sense of learning from others' mistakes.

Forecasting errors

These cases are recent expensive mistakes in the area of forecasting. They give a guide as to what can and does go wrong.

CHARTERING OIL TANKERS

In his book *Practical Experiences with Modelling and Forecasting Time Series* Gwilym Jenkins cites the case of an oil company that lost a lot of money by making a superficial analysis of a time series. Figure 7.1 shows the spot prices for chartering oil tankers for the years 1968–71. Noticing an upwards trend in the data in 1969 and early 1970 the company assumed that the trend would continue. Long-term contracts were taken out for future chartering. Since the company was large, this affected the market mechanism and prices were pushed artificially high. This prompted more long-term chartering and as a result further artificial increases. When the long-term chartering stopped, the spot price fell to its early 1970 level.

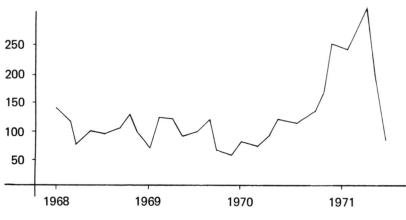

Figure 7.1

A deeper analysis of the data 1968–70 would have revealed that the series was a 'random walk'. This means that the step from one value to the next is a random one. Since each movement is random there is no pattern, or trend, in the data that can be used to make forecasts.

Gwilym Jenkins suggests that in this sort of situation it would have been better to investigate that factors causing the price for chartering oil tankers to vary; factors such as supply of tankers, demand for oil, etc. Two mistakes were made in this case. First, it was not recognized that a large company can itself have a big effect on prices in a market which, in the short term, is of fixed size. Secondly, the wrong type of forecasting technique was used. It has been estimated that the cost of these mistakes was of the order of hundreds of millions of pounds sterling.

SALES OF CHILDREN'S CLOTHING

A director of a clothing company noticed a numerical link between sales of children's clothing and the number of new car registrations. The company developed a regression model to demonstrate the link between the two variables. The analysis showed the link was a strong one with a high correlation coefficient and the model was used for planning the company's future business. However, the correlation demonstrated only that there was *association* between the variables, it did not prove that the link was a *causal* one. Over the period of the analysis both variables had risen steadily. Soon the economic situation changed and there was a decrease in new car registrations. Very poor forecasts were produced. The analysis had shown only that there was association between the variables during the set of circumstances which applied in the past. To extrapolate and use the model in a set of circumstances (decreasing new car registrations) for which the model had not been proved, made it very likely, in the absence of a causal link, that poor forecasts would be generated.

There is an additional problem in this case. In order to forecast sales the company had first to forecast new car registrations, which proved just as difficult as forecasting sales of clothing. A more realistic approach might be to relate sales of children's clothing to (a) the number of births an appropriate number of years previously, to reflect the number of children for whom clothes could be bought; (b) a macroeconomic variable reflecting the money in consumers' hands.

PRODUCTION PLANNING IN THE MANUFACTURING INDUSTRY

Forecasts are not always monitored for accuracy after the event. A company in the manufacturing industry did check sales forecasting accuracy and found that the predictions were excellent. However, production planning for which the forecasts had been prepared was poor. After investigation it was found that the forecasts, delivered each week in the form of a large and heavy computer print-out, were never used. The production planners could not understand the output and so ignored it and used their own judgemental forecasts, which were poor. The forecasters had made no attempt to produce their forecasts in a form the users could understand. The forecasting had never been integrated with the decision-making it was intended to serve.

Stories of errors in forecasting are usually of inaccurate forecasts being used, with disastrous results. But it may be that a more frequent situation is as in the above case, where accurate forecasts are not used, with disastrous results.

The above three cases illustrate that while forecasting errors do occur because of technical mistakes, the errors caused by companies not properly organizing the ways in which the forecasts are used (non-technical, management mistakes) might be far more important. The next section looks at this problem and discusses how forecasting can be integrated into a company's management processes.

Incorporating forecasting into a company's management systems

The key word here is systems. Forecasting should not be viewed as a number-generating technique but as a system. This broader view makes it easier to see that the technique is just one part of the fore-casting process which should include ways of introducing qualitative judgements into the forecasts, of monitoring the effectiveness of the process and of adjusting the system in response to feedback. In addition it leads to considering the links the forecasting system has with other management systems within the organization. The lack of real thought about the nature of the links with other management systems is often the cause of forecasts being wrongly used.

Gwilym Jenkins suggests some guidelines for the development of a forecasting system.

(1) *Analyse decision-taking systems* served by the forecasts. Set

down all decisions and actions that will be influenced by the forecasts. For instance, forecasts of car sales might be required by the manager of an assembly line at a car plant. Primarily those forecasts will help decide the speed of the line; but other decisions will be influenced by the forecasts: the ordering of steel, the production of sub-assemblies, the buying of components, are examples. The forecaster should know all the decisions that have some connection with his forecasts.

(2) *Define* what forecasts are needed. This can only be done when the decision-making system has been investigated. It comprises determining forecast variables, time horizons and frequencies, accuracy levels, etc. It ensures that needless forecasts are not generated and that decisions which require a forecasting input are properly supplied. In the car assembly example the exact nature of the forecasts would be determined after consideration of all the decisions affected.

(3) *Develop a conceptual model* of the forecasting method, including a list of those factors which may influence the variable being forecast. This suggests what the ideal forecasting method might be. It also helps prevent the use of associative models under the assumption that they are causal.

(4) *Ascertain the data that are available* (and those not available) to indicate how the actual forecasting method might fall short of the ideal.

(5) *Develop the actual method* for making forecasts. This is the 'technique' part of the system that is often the only part given any real consideration. The step involves the selection of a suitable technique based upon the type of forecasts required, the accuracy required and the data and resources available to the organization.

(6) *Test the method*, to see how accurate the forecasts are likely to be. This step is described in more detail below.

(7) *Decide how judgements* should be incorporated into the forecasts. Quantitative forecasting models work on the assumption (not always explicitly recognized) that many of the conditions of the past will continue in the future. Other factors, for example political circumstances, may result in future conditions being radically different from the past. Views about such changes should be incorporated into the forecast. This must not be done by making instinctive or arbitrary changes. A systematic method for incorporating judgements should be developed. This may lean on one of the qualitative forecasting

techniques described later. It will certainly require forecasters to be accountable for the changes they make.

(8) *Implement* the forecasting system. This means taking trouble to ensure that the newly developed forecasting system is properly linked with the existing management systems. When the system is first used the forecaster should be available for advice, and to verify that its use is understood correctly.

(9) *Monitor* the performance of the forecasting system. The way the system is operating should be continually checked to see that things are happening as specified, both in the use being made of the system and in the system's statistical performance. Tests of its accuracy should be made with a view to changing the technical structure of the model as conditions change and improving forecasting accuracy (rather than allocating blame).

To reiterate, the major point being made in discussing forecasting as a system is that forecasting seems to fail within an organization far more often because the broader management issues have been neglected than because of technical matters. The second worked example at the end of the chapter shows how the guidelines might be used.

Review of forecasting techniques

Forecasting techniques can be divided into three categories: qualitative; causal modelling; time series modelling. This review describes in outline a range of methods, but the list is not intended to be exhaustive.

QUALITATIVE METHODS

Contrary to widely held opinion, the few tests that have been done show that judgemental or qualitative forecasting methods are less accurate than quantitative ones. Part of the reason for this may be that qualitative forecasts should be made just as carefully and systematically as quantitative ones, whereas in practice they are made haphazardly. The better of the methods described below are systematic ways of bringing together different judgements into a forecast.

Historical analogy

The variable being forecast is assumed to behave in the same way as a similar variable in the past. For instance, with the growth of sales of a new product it is forecast that sales will grow just as they did for an existing product which is deemed to be similar. Provided there are sufficient potential analogies to choose from, evidence suggests this method is quite accurate. For example, the producers of a new and very successful alcoholic beverage predicted its sales through consideration of the sales patterns of other similar products which had been successful.

Panel consensus

A panel of 'experts' debate the factors affecting the variable and try to agree on a forecast. Apart from a one-man guess this is perhaps the most common method of forecasting. It is also one of the least accurate. It suffers from the disadvantage that group effects sway the outcome. The forecast is affected as much by personalities and rank as it is by knowledge.

Delphi

This is a variant on panel consensus which seeks to eliminate the disadvantages of that method. The panel are kept apart, either by being in different places or by only being allowed to communicate through the chairman. Group effects should not therefore sway the decision. The chairman asks each panel member to make his forecast for the variable together with his reasons, without other panel members knowing what he proposes. The chairman collects these statements and gives the panel feedback on (a) the average forecast, (b) the range and (c) the principal reasons. The panel are then asked to make a second forecast after reflecting upon the feedback. The process continues until some measure of agreement is reached. Tests show the method to be accurate. It is also convenient since it can be carried out by telephone or post.

Market research

Market research is well known. It has been included for completeness. The essence of this method is the collection of a large sample of views about, say, a new product, and their being averaged out. If the sample has been collected properly, the 'swings-and-roundabouts' effect should eliminate biases. It is, usually, expensive to carry out.

CAUSAL MODELLING

Causal modelling means that the variable to be forecast is related, usually by regression analysis, to one or more other variables which it is thought in some way, 'cause' changes in the forecast variable. Sometimes the causal mechanism is definite; other times it is assumed. The methods are described in outline only since the statistical method on which they are based is regression analysis which was described in the last chapter.

Explanatory models

If there is assumed to be a mechanism whereby a series of 'explanatory' variables influence the variable to be forecast, then multiple regression analysis can be used to estimate the relationship more exactly. For example, 'explanatory' variables such as advertising, price and last year's sales may be used to predict this year's sales. Even when the relationship is not a linear one, variations on multiple regression can sometimes be used to estimate it.

Leading indicators

In the above, judgements had to be made as to which variables could be said to 'explain' the variable to be forecast. In some situations there exist variables whose movements precede those of the forecast variable and these leading indicators can be incorporated into the model. For example, a leading indicator of sales of 4-year-old children's clothing might be the birth rate 4 years previously. A leading indicator is handled just as another explanatory variable in a regression analysis.

Econometric models

Econometric models are sets of equations describing, mathematically, relationships between economic variables. The difference between econometric models and other causal models is that they deal exclusively with economic variables. Usually they are large, comprising many equations. The variables being forecast appear in some equations as dependent variables (y variables) and in others as independent variables (x variables). More complicated procedures have to be used to facilitate the simultaneous estimation of coefficients in different equations.

TIME-SERIES METHODS

Time-series methods are ways of predicting future values of a variable solely from historical values of itself. They are useful because (a) they provide good forecasts for those variables that are stable, and therefore where past conditions will continue into the future; (b) they provide good short-term forecasts where there is not enough time for conditions to change substantially; and (c) even where conditions are changing they provide a base from which to incorporate the changes into the forecast. The techniques described are categorized according to the types of series to which they are applicable.

Stationary series

A data series is stationary if it fluctuates about some constant level and, while the amount of fluctuation differs from one time period to the next, there is no general tendency for there to be more fluctuation at one part of the series than at another. More technically a stationary series has no trend and has constant variance. A trend is some consistent movement upwards or downwards.

In the long run, very few series are stationary, but they may be in the short run. For example, the weekly stock volumes in a warehouse over 2 years is a long series (104 observations) which may well be stationary. Over 5 years it would probably not be.

(i) *Moving averages*
The original series is replaced by a 'smoothed' series, obtained by replacing each actual observation with the average of it and other observations either side of it. If each average is calculated from 3 actual observations, it is said to be a 3-point moving average; if each is from 5 actual observations, it is a 5-point moving average, and so on.

Example: Table 7.1 shows a 3-point moving average.

Table 7.1

Time period (t)	Actual series (x)	Smoothed series
1	17	
2	23	17.3 = (17 + 23 + 12)/3
3	12	16.3 = (23 + 12 + 14)/3
4	14	17.3 = (12 + 14 + 26)/3
5	26	18.7 = (14 + 26 + 16)/3
6	16	

Smooth value at time period t

$$= \frac{\text{Actual value at } t - 1 + \text{Actual value at } t + \text{Actual value at } t + 1}{3}$$

i.e. $\quad S_t = \dfrac{x_{t-1} + x_t + x_{t+1}}{3}$

The averaging process is intended to smooth away the random fluctuations in the series. The forecast for any future time period is the most recent smoothed value. In the series in Table 7.1 the forecast is 18.7 for all periods in the future. A constant forecast makes sense because of the stationarity of the series.

Seasonal as well as random fluctuations in the data can be smoothed away by including sufficient points in the average to cover the seasonality. Seasonal monthly data would be smoothed using a 12-point moving average. Each month is included once and only once in the average, and thus seasonal variation will be averaged out.

The use of an even number of points in a moving average creates a problem. A smoothed value no longer can refer to a particular time period. It must refer to halfway between the middle two time periods. For example, a 3-point moving average for the months January, February, March is the smooth value for February. A 4-point moving average for January, February, March, April is the smooth value for in between February and March. This makes no difference when forecasting a stationary series, but it does have an effect in other uses of moving averages.

The decision as to how many points to include in the moving average is based on the seasonality of the data. In the absence of seasonality the average should include sufficient points to be able to smooth the fluctuations, but not so many that the last smoothed value refers to a time period remote from the time periods for which the forecasts are being made. In practice, 3- or 5-point moving averages are probably the most common.

Even when the data are non-stationary, the method of moving averages can still be used to smooth out random fluctuations, enabling trends and other patterns to be seen more clearly.

(ii) *Exponential smoothing*

For a moving average, each value in the average was given an equal weight. In a 3-point moving average each value is given a weight of 1/3. Exponential smoothing is a way of constructing an 'average' which gives more weight to recent values of the variable.

The smoothed series is given by the equation

new smoothed value

$$= (1 - \alpha) \text{ (Previous Smoothed value)}$$
$$+ \alpha \text{ (Most recent actual value)}$$

i.e. $S_t = (1 - \alpha) S_{t-1} + \alpha x_t$

where α is between 0 and 1.

The value of α (said alpha) is chosen by the forecaster. The larger its value, the heavier the weighting being given to the recent values. Its value may be selected after testing out several values by the methods described in a later section. In practice α is usually in the range 0.1 to 0.4.

Example. The same data used in table 7.1 are exponentially smoothed in table 7.2. Since the smoothing equation requires a previous smoothed value to get it started, it is usual to make the first smoothed value equal to the actual value. This assumption will have a negligible effect unless the series is a very short one.

Figure 7.2 shows how exponential smoothing works to average out random fluctuations. As with moving averages the forecast for future time periods of a stationary series is the most recent smoothed value, in this case 17.84.

Technical note. A little algebraic manipulation is required to show

Table 7.2

Original series	Smoothed series (using α = 0.2)
17	17
23	18.2(= 0.8 x 17 + 0.2 x 23)
12	16.96(= 0.8 x 18.2 + 0.2 x 12)
14	16.37(= 0.8 x 16.96 + 0.2 x 14)
26	18.30(= 0.8 x 16.37 + 0.2 x 26)
16	17.84(= 0.8 x 18.30 + 0.2 x 16)

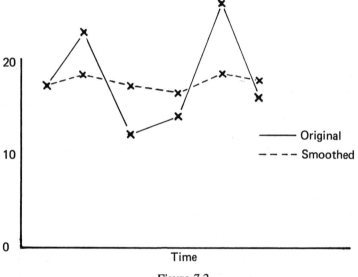

Figure 7.2

why exponential smoothing gives different weighting to different time periods. The equation for exponential smoothing is:

$$S_t = (1 - \alpha) S_{t-1} + \alpha x_t$$

But, from the previous time period,

$$S_{t-1} = (1 - \alpha) S_{t-2} + \alpha x_{t-1}$$

Putting this S_{t-1} in the top equation gives

$$S_t = (1 - \alpha)^2 S_{t-2} + (1 - \alpha) \alpha x_{t-1} + \alpha x_t$$

Just as S_{t-1} in the original equation was substituted, so S_{t-2} can be substituted. Continuing this process eventually gives:

$$S_t = \alpha x_t + \alpha(1 - \alpha)x_{t-1} + \alpha(1 - \alpha)^2 x_{t-2} + \alpha(1 - \alpha)^3 x_{t-3} + \ldots$$

The weightings being given to past values are

$$\alpha, \alpha(1 - \alpha), \alpha(1 - \alpha)^2, \alpha(1 - \alpha)^3 \ldots$$

Since α, and thus $1 - \alpha$, lie between 0 and 1, these weightings are decreasing. For instance if $\alpha = 0.2$, the weightings are,

$$0.2, 0.16, 0.128, 0.1024, 0.0819 \ldots$$

Recent actual values receive heavier weighting than earlier ones. The smoothing equation derived above illustrates how the weighting works. It is not intended to be used for calculations.

Series with a trend

The use of moving averages or exponential smoothing may reveal the existence of a trend; or the trend may have been immediately obvious without any smoothing. For a non-stationary series the techniques have to be adapted before they can be used. There are several variants of moving averages and exponential smoothing that can deal with a trend. The one described here is Holt's method which is a form of exponential smoothing.

Holt's method
The formula for exponential smoothing is

$$S_t = (1 - \alpha)S_{t-1} + \alpha x_t$$

If the series has a trend, the smoothed value S_t (which is the forecast for future time periods) will generally be too low since (a) it is formed in part from the previous smoothed value S_{t-1} and (b) the forecast does not allow for the effect of a trend on future values. If there is a trend, it should be seen in the smoothed values, therefore a first way of calculating a trend might be:

Trend = Most recent smoothed value − Previous smoothed value

i.e.   Trend at time $t = S_t - S_{t-1}$

Just as random fluctuations in the actual data can be smoothed, so it is with the trend. A smoothed estimate of the trend is obtained by using a smoothing constant (labelled γ, said gamma) to combine the most recently observed trend ($S_t - S_{t-1}$) with the previous smoothed trend. γ is between 0 and 1, is chosen by the forecaster and may or may not be different from α.

Smoothed trend = $(1 - \gamma)$ Previous smoothed trend
$+ \gamma \cdot$ Most recently observed trend

i.e. $b_t = (1 - \gamma)b_{t-1} + \gamma(S_t - S_{t-1})$

How is this estimate of the trend used in conjunction with the exponential smoothing formula? Firstly the basic formula is changed so that the previous smoothed value, S_{t-1} is increased (or decreased) to allow for the trend.

$$S_t = (1 - \alpha)S_{t-1} + \alpha x_t$$

becomes

$$S_t = (1 - \alpha)(S_{t-1} + b_{t-1}) + \alpha x_t$$

Secondly, future forecast values allow for the effect of the trend. A forecast for 3 periods ahead is no longer S_t, but:

Forecast 3 periods ahead = Most recent smoothed value +3 x Trend

More generally, the forecast for m periods ahead, F_{t+m}, is given by:

$$F_{t+m} = S_t + m \cdot b_t$$

To summarize, when a time series has a trend, forecasts with Holt's method are based on 3 equations.

$$S_t = (1 - \alpha)(S_{t-1} + b_{t-1}) + \alpha x_t$$
$$b_t = (1 - \gamma)b_{t-1} + \gamma(S_t - S_{t-1})$$
$$F_{t+m} = S_t + m \cdot b_t$$

where x_t = actual observation at time t;
S_t = smoothed value at time t;
α, γ = smoothing constants between 0 and 1;
b_t = smoothed trend at time t;
F_{t+m} = forecast for m periods ahead.

Example. Table 7.3 shows how Holt's method is applied to an annual series of sales figures. The series has been shortened in order to simplify the example. The smoothing constants have values:

$$\alpha = 0.2; \quad \gamma = 0.3$$

Table 7.3

Year	Sales volume	$\alpha = 0.2$ Smoothed sales	$\gamma = 0.3$ Smoothed trend
t	x_t	s_t	b_t
1975	12	12.0	—
1976	15	15.0	3.00
1977	20	$18.4 = 0.8(15 + 3) + 0.2 \times 20$	$3.12 = 0.7(3.00) + 0.3(18.4 - 15)$
1978	21	$21.4 = 0.8(18.4 + 3.12) + 0.2 \times 21$	$3.08 = 0.7(3.12) + 0.3(21.4 - 18.4)$
1979	25	$24.6 = 0.8(21.4 + 3.08) + 0.2 \times 25$	$3.12 = 0.7(3.08) + 0.3(24.6 - 21.4)$
1980	28	$27.8 = 0.8(24.6 + 3.12) + 0.2 \times 28$	$3.14 = 0.7(3.12) + 0.3(27.8 - 24.6)$
		Forecasts	
1981		$30.94 = 27.8 + 3.14$	
1982		$34.08 = 27.8 + 2 \times 3.14$	
1983		$37.22 = 27.8 + 3 \times 3.14$	

The choice of smoothing constants is based on the same principle as for ordinary exponential smoothing.

The calculating process needs a starting point both for the trend and the smoothed values. The smoothed values for the first 2 time periods are taken to be equal to the actual values. The trend cannot be calculated from just the first time period. The smoothed trend for the second time period is taken to be equal to the difference between the first 2 actual values.

Series with trend, seasonality and cycles

A *cycle* is some regular repeating pattern of upwards and downwards movements in a time series. For example, economists discuss the existence of 7-year cycles in economic variables. *Seasonality* is cycles that are of length no greater than 1 year. For example, sales of heating fuel have a seasonal pattern.

There are techniques that can deal with series exhibiting cycles and seasonality. Many are based on principles such as moving averages and regression. However, they tend to be long and complex. The bibliography gives references which describe the techniques in detail.

Assessing the accuracy of a forecasting method — before it is used

Item (6) in the guidelines for a forecasting system (p. 192) concerned testing the accuracy of a forecasting method in advance. Two ways of doing this will be presented. The first is for smoothing techniques only. The second is for any technique, but is particularly used for techniques based on regression analysis.

SMOOTHING METHODS

The test involves comparing each one-period-ahead forecast with the actual observation for that time period. A measure of scatter for the differences is calculated. The method which has the lowest scatter of the differences is the best.

For example, table 7.4 reproduces the example used for moving averages (table 7.1). The smoothed figures are based on 3-point moving averages. The first smoothed figure is 17.3 calculated on the actual data for time periods 1–3. It is the forecast for time periods 4 and onwards. The difference column is the comparison between each

Table 7.4

Time period	Actual data	Smoothed data	Forecast	Difference
1	17			
2	23	17.3		
3	12	16.3		
4	14	17.3	17.3	−3.3
5	26	18.7	16.3	9.7
6	16		17.3	−1.3

forecast and the first period for which it is a forecast. The scatter of the difference is usually measured by the mean square error (MSE). This is similar to the variance but uses n instead of n − 1 in the denominator

$$\text{MSE} = \frac{(-3.3)^2 + (9.7)^2 + (-1.3)^2}{3}$$

$$= 35.6$$

Different smoothing methods can be compared using the MSE for each, calculated, of course, over the same time periods. The mean absolute deviation (MAD) may be used instead of the MSE.

Just as different smoothing techniques can be compared by this process, so the same exponential smoothing technique but with different smoothing constants can be evaluated to select the best constant.

OTHER METHODS

The above method of testing cannot be used for the regression methods since they do not accumulate forecasts as they go along, but use the whole of the available data at once. Figure 7.3 shows some time series sales data. Forecasts are based on calculations using all the data available up to point A. To test the accuracy of the method (and to compare it with other methods including smoothing methods), forecasts are made for the period from B to A using only the data up to point B.

Forecast and actual over this period are compared, and the MSE or

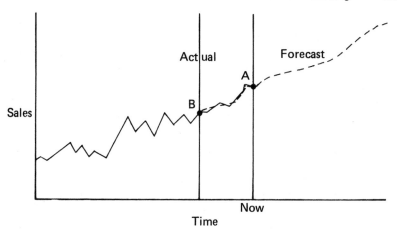

Figure 7.3

MAD calculated. The method with the best forecasts (lowest MSE) over the period B to A would be chosen. For forecasting the future all the data up to A would be used.

The difference in application between these two methods for testing accuracy is that the former is for smoothing methods only, whereas the latter can be used for any method including smoothing methods.

Assessing the accuracy of a forecasting method — after it has been used

The first tests that can be made are statistical ones. Are the differences between the forecasts generated and what actually happened random, or is there a pattern in them? How large are the differences? If they are random, then, even if they are large, the forecasting system has nevertheless probably done as well as can be expected. The time series just happens to contain a large random, and therefore unforecastable, element. If there is a pattern in the forecast errors then the pattern should be analysed and incorporated into the model. In other words if a recognized pattern is seen outside the model, then the model is incomplete and the pattern should be brought into the model. Randomness can be tested visually or by a variety of statistical tests. As before, the size of the errors is calculated through a measure of scatter, typically the mean absolute deviation or mean square error.

If one takes a broader view of forecasting, however, one can see that purely statistical tests on the closeness of the forecasts to actuality are inadequate. Forecasts are made to enable managers to take action. The actions may change the basis on which the forecasts were made and render them inaccurate. For instance, if forecasts of future sales suggest they are to be substantially below the company's objective, then management may take action to boost sales. If the action is successful the original forecasts will be very different from actual sales, and this could be interpreted as poor forecasting when in fact the forecasting system has done its job effectively and successfully.

Another qualitative aspect of measuring the success of a forecasting system concerns the extent to which forecasting methods should be expected to predict very rare chance events. A man who forecast his personal cashflow over the next year may find his forecasts very wrong if he wins a large amount on the football pools. The forecast, although wrong, may still have been a good one, because it is not possible for anyone or any method to forecast a pools win. Freak, random events should generally be excluded when assessing the accuracy of a forecasting system unless, of course, it was the purpose of the system to predict just such events.

In general, then, one should be aware of qualitative as well as statistical factors when assessing the accuracy of a forecasting method.

Worked examples

1. SOUTHERN CALIFORNIA RESEARCH AND PRODUCTION (ELECTRICAL)

The production manager has to prepare a forecast of the demand for toasters next month. The data shown in table 7.5 are available. It is assumed that there is no trend.

Table 7.5

1981			1982							
Oct	Nov	Dec	Jan	Feb	Mar	Apr	May	June	July	Aug
2000	1350	1950	1975	3100	1750	1550	1300	2200	2775	2350

(a) Use a 3-month moving average to compute a forecast of monthly demand for toasters.

(b) Use exponential smoothing with α = 0.1 to compute forecasts.
(c) Compare the accuracy of the 2 forecasting methods using whatever measure you think appropriate.
(d) Which of the two methods works best? Why do you think this is so?
(e) Make a point forecast for September 1982.

(a)

		Demand	3-month moving average	Forecast	Deviation
1981	Oct	2000			
	Nov	1350	1770		
	Dec	1950	1760		
1982	Jan	1975	2340	1770	205
	Feb	3100	2275	1760	1340
	Mar	1750	2130	2340	−590
	Apr	1550	1530	2275	−725
	May	1300	1680	2130	−830
	June	2200	2090	1530	670
	July	2775	2440	1680	1095
	Aug	2350		2090	260

This assumes that the data show no trend, or, if there is a trend, it is not sufficiently large to affect short-term forecasts over a period of, say, 3 months.

(b)

		Demand	Exponential smoothing (α = 0.1)	Forecast	Deviation
1981	Oct	2000	2000		
	Nov	1350	1935		
	Dec	1950	1936		
1982	Jan	1975	1940	1936	39
	Feb	3100	2056	1940	1160
	Mar	1750	2025	2056	−306
	Apr	1550	1978	2025	−475
	May	1300	1910	1978	−678
	June	2200	1939	1910	290
	July	2775	2023	1939	836
	Aug	2350	2056	2023	327

The same assumptions apply as with the moving average case.

(c) MSE for exponential smoothing

$$= \frac{\text{Sum of squared deviations}}{\text{No. of months for which forecast compared with actual}}$$

$$= \frac{39^2 + 1160^2 + (-306)^2 + (-475)^2 + (-678)^2 + 290^2 + 836^2 + 327^2}{8}$$

$$= \underline{376,999}$$

MSE for moving average

$$= \frac{\text{Sum of squared deviations}}{\text{No. of months for which forecast compared with actual}}$$

$$= \frac{205^2 + 1340^2 + (-590)^2 + (-725)^2 + (-830)^2 + 670^2 + 1095^2 + 260^2}{8}$$

$$= \underline{639,472}$$

(d) Exponential smoothing has the lower MSE and therefore performs better over the time series. However, the series is so volatile in such an apparently random fashion, that it would be difficult for any time series method to do well. It may be better to search for the reasons for the volatility and try to develop a causal model.

(e) Since exponential smoothing performs better, the forecast made by this method for September 1982 would be chosen, 2056.

2. ARTS CENTRE THEATRE

The theatre organization stages productions from Shakespeare to contemporary experimental ones. Some productions are booked up long before the first night; others are less than half-full. In planning future activities the organization has asked a firm of consultants to advise how forecasts of the demand for each production might be prepared. Briefly describe how the guidelines for a forecasting system might be applied to this situation.

The guidelines give a checklist to aid the preparation of a forecasting system. Their purpose is to enable a broad view of the problem to be taken. There are nine stages to the guidelines. What follows is a brief description of how they might be applied to the theatre organization.

(1) *Analyse the decision-taking system.* In planning future productions many interlinked decisions will have to be taken, all of which depend in some way on the forecasts:

- length of run
- seat prices
- discounts
- promotions
- advertising
- costs
- and many more.

The timing of these decisions, the people who take them, and how decisions are altered or reviewed should be mapped out. In the light of knowledge about how these decisions are taken at present, it may be necessary to suggest changes to the decision-making process.

(2) *Define forecasts needed.* Having completed stage 1, it should be possible to see what forecasts of which level of accuracy are needed, when they are needed, and how, and by whom they will be used.

(3) *Prepare a conceptual model.* The factors which are likely to affect a forecast of demand for theatre tickets should be considered and listed. The main factors are probably:

(a) Internal: reputation of the play; reputation of the actors/the presence of star names; ticket prices; advertising/promotional expenditure; demand for previous, similar productions.
(b) Environmental: economic situation; weather/time of year; day of week; rival attractions.

These are the factors that it would be hoped to take into account in any forecasting model.

(4) *Find out what data are available.* It is possible that all data in connection with the above factors are available. More usually there will be gaps which, if they cannot be filled, will constrain the choice of techniques. In this case it is particularly likely that while attendance figures are available, demand is not. This is particularly important because of the 'house full' productions where demand may well have

been 3 or 4 times the attendance. Since it is demand that is to be forecast some subjective assessment of the difference between demand and attendance may have to be made. Equally, advertising expenditure may not be available for individual productions and judgement may be required to break it down.

(5) *Decide on the technique to be used.* Knowing the forecast required and the data available the forecasting technique can be decided upon. Here it is almost certain that a causal model would be used relating demand to some of the factors listed in (3) above.

(6) *Test the accuracy.* Supposing a causal model had been chosen, the accuracy would be tested in two ways. First, the correlation coefficient would measure the closeness of fit of the model to actual data. Secondly, data from perhaps two productions would be put to one side. The coefficients of the causal model would be estimated using the remaining data, and this model used to forecast demand for seats at the two productions. The performance of the model would be quantified by a measure of scatter (MSE) for the deviations between forecast and actual.

(7) *Decide how to incorporate judgement.* Situations will arise where special circumstances warrant the making of adjustments to a statistical forecast. Publicity prior to the opening of a play, or the withdrawal of a famous actor, are examples of non-quantifiable influences on demand. Judgement is needed to modify the forecasts. A system, perhaps regular monthly meetings, should be formalized whereby alterations can be discussed and those making the alterations made accountable for them. It is essential that any changes should be carefully considered by all involved rather than being the results of unilateral and hasty decisions.

(8) *Implement.* The manager of the forecasting system should follow the first set of forecasts through the system. He already knows (stage 1) who is using them and for what decisions. He must now make sure they are being used as intended and find out in what ways, if any, they are inadequate.

(9) *Monitor.* Implementation refers to the first use of the forecasts. In monitoring, the performance of the system is watched, but not in such detail, every time it is used. The accuracy of forecasted demand for each production should be recorded and reasons for deviations explored. In the light of this information the usefulness of the system can be assessed and changes to it recommended.

Table 7.6   Comparison of forecasting methods

Category	Techniques	Comment
Qualitative Used in new situations or when data are scarce	Analogy	Can only be used when suitable analogies are available
	Panel	Subject to rank/personality influences
	Delphi	Overcomes disadvantages of the Panel method
	Market research	Expensive, but usually accurate
Causal Based on relating forecast variables to other variables; usually based on regression and the reservations about regression apply	Explanatory	When other variables 'cause' changes in the variable to be forecast
	Leading indicator	When other variables 'lead' the forecast variable
	Econometric	Economic variables only
Time series Used in stable situations or as a base forecast	Moving average Exponential smoothing	For forecasting stationary series, or to spot trends/patterns in data
	Holt	For series with a trend

Final comments

In this chapter the management aspects of forecasting have been stressed, since this is an area of expertise that has been lacking in applied forecasting. Forecasting techniques have been reviewed, but as so many are available, only a small fraction could be covered. However, these are sufficient for a start to be made on most sets of data. The bibliography indicates where details of other techniques may be found. Table 7.6 summarizes the techniques reviewed.

It is probably true that sensible people should only use good forecasts, not try to make them. The general public, and the business world, judges forecasts very harshly. Unless they are exactly right they are failures; and, of course, they are never exactly right. This black or white test of forecasting is unfortunate. The real test is whether the forecasting is, on average, better than the (possibly) crystal-ball alternative.

A more positive view is that the last quarter of this century is particularly rewarding time to try forecasting. The volatility of data series referred to at the outset of the chapter puts a premium on good forecasting. At the same time the facilities to make forecasts are suddenly readily available in the form of sophisticated calculators and mini-computers. The use of a simple technique takes only a little time and can make a considerable difference to planning.

Further examples

1. UNITED AND GENERAL HAMBURGERS

Table 7.7 shows the monthly sales from a fast-food chain for an 18-month period from March 1981 to August 1982. Forecast sales for the 3 months September–November 1982. There is a trend which must be allowed for. Because of this, and because there are insufficient data to establish cyclical and seasonal effects, it is suggested that Holt's method be used with $\alpha = 0.2$ and $\gamma = 0.3$.

2. UNITED AND GENERAL HAMBURGERS (continued)

In the above example, would it have been preferable to choose $\alpha = 0.1$, $\gamma = 0.1$? Make the decision by testing the accuracy of each set of values in the way suggested for smoothing methods.

Table 7.7

		£ Thousand
1981	Mar	226
	Apr	242
	May	257
	June	283
	July	284
	Aug	278
	Sep	297
	Oct	317
	Nov	321
	Dec	330
1982	Jan	350
	Feb	362
	Mar	377
	Apr	395
	May	396
	June	407
	July	408
	Aug	418

ANSWERS

(1) *United and General Hamburgers*

1982	Sept	436.1
	Oct	446.0
	Nov	455.8

(2) *United and General Hamburgers*

$\alpha = 0.2$, $\gamma = 0.3$; MSE = 145.0
$\alpha = 0.1$, $\gamma = 0.1$; MSE = 546.5

The mean square errors cover the 15 months from June 1981 to Aug 1982.

Decision-making Techniques

Part IV describes three techniques that can be applied to decision problems. The techniques are *Decision analysis, linear programming* and *simulation*. They should not be regarded as decision-takers in themselves but as ways of handling data to aid the decision-taking. Whilst most decision problems have some quantitative aspects, few are exclusively quantitative. The techniques can suggest the best decision given the data available, but other, qualitative factors may be involved. The techniques should therefore be regarded as providing a sound analytical base from which to take decisions. There is room for both judgement and analysis.

Preparation

The technical content of Part IV is, happily, less than that of Part III. Chapter 8 requires knowledge of probability. Chapter 9 is the more difficult technically, involving simultaneous equations (Appendix E). There are no prerequisites for chapter 10.

CHAPTER 8

Decision analysis

By the end of the chapter the reader should know how decision analysis works and when it can be applied. The chapter contains sufficient information to enable the technique to be applied to most problems. The basic technique can be extended in terms of providing extra information and of calculating probabilities more accurately.

Decision analysis is a technique which acts as an aid to the solution of a particular type of decision problem. All decision problems have two common elements. First, there is a decision to be taken, which might be the setting of a variable at a specific level; for example: how much to invest in a capital project? Or it might be a yes/no decision such as: should the company expand production or not? Or it might be qualitative decision; for example: what should be the range of colours for a new model of car. Second there is a criterion by which one can evaluate the outcome of the decision. In business, the criterion is often profit. Beyond these two common elements the decision problem may have special characteristics which categorize it as being of a particular type, and which enable it to be solved by an appropriate technique.

A primary difficulty is matching the decision problem with the suitable technique. In order to do this for decision analysis, it is necessary to consider what the characteristics of a decision problem are which make it suitable for solution by this technique.

Characteristics of decision analysis problems

(1) The problem comprises a series of sequential decisions interwoven with chance events spread over time. For instance, the problem of launching a new product has several stages.

Decision	what market research to undertake.
Chance event	result of market research.
Decision	whether to have a preliminary launch in a test market.
Chance event	result of test market.
Decision	whether to launch or abandon.
Chance event	success or failure of product.

This decision/chance event/decision/chance event, etc. pattern is the single most important characteristic of decision problems that are solvable by the technique of decision analysis.

(2) The decisions within the series are not quantitively complex. The decisions in the product launch example are of the type: choose between a route which will result in a likely total profit X and a route which will result in a likely total profit Y. The choice is then to pick the larger of the two numbers X and Y. The decision is therefore quantitatively simple (although it is complex in other senses). Compare the question of how to distribute most efficiently a manufactured product throughout the UK. As formulated, this comprises one once-for-all decision based on all the information that is ever going to be available. There is one decision and no chance events. However, the one decision hinges on very complex calculations. This type of problem is not suitable for the application of decision analysis.

(3) There are other less important characteristics which will emerge during the chapter.

Prerequisite ideas

Before looking at decision analysis it is necessary to understand a number of concepts and terms which are part of it. To describe these concepts a simple example will be used: that of deciding whether to accept the following offer. An unbiased coin will be tossed. If the coin falls 'heads' one receives £5, if it falls 'tails' one loses £4. The example is trivial, but it has a value in describing basic concepts.

DECISION TREE

The logical sequence of decisions/chance events when represented diagrammatically is called a decision tree:

 represents a point at which a decision is taken (called a *decision node*);

○ represents a point at which a chance event occurs (called a *chance event node*);

—— represents the logical sequence between nodes.

In the coin example, the decision tree is as shown in figure 8.1.

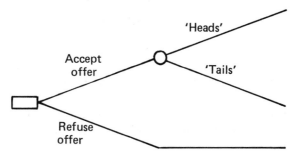

Figure 8.1

PAYOFFS

At the end of each 'branch' of the decision tree there is a payoff (usually in monetary terms, but it can be non-monetary) which is the result of following a path through the tree. In figure 8.2 the payoffs are shown at the end of the branches.

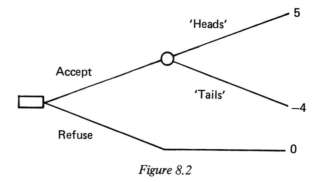

Figure 8.2

EXPECTED MONETARY VALUE (EMV)

Every decision and chance event has associated with it an EMV. This is defined as the average benefit which would accrue from a decision or chance event if it occurred many, many times. In the example, if the coin were tossed 100 times, the most likely result (since the coin is not biased) is 50 heads and 50 tails.

$$\text{Total profit} = (50 \times 5) - (50 \times 4)$$
$$= 250 - 200$$
$$= £50$$
$$\text{Average profit} = \frac{50}{100} = 0.5 \ (= 50\text{p})$$

Since probability can be defined via frequency of occurrence, a more convenient calculation of EMV is to multiply probabilities by payoffs and add.

$$\text{EMV} = (0.5 \times 5) - (0.5 \times 4) = 0.5$$

The EMV at the chance node is therefore 50p, and consequently the EMV associated with the accept decision is 50p. The alternative decision 'refuse' has of course an EMV of 0. One chooses the decision with the higher EMV.

EMV is typically used as the criterion for decision-taking. It is a popular criterion, but it is not the only one. For example, the payoffs and the criterion could be measured in terms of expected opportunity loss. Other criteria will be discussed briefly at the end of the chapter.

Applications

TECHNOLOGICAL DEVELOPMENTS

A project may require several technological developments in order to complete it. Concorde, and North Sea gas and oil, are examples. The exact nature of these developments is not known at the outset of the project. To achieve a breakthrough in one area, several channels of research may be pursued. The channel with the most successful outcome is the one incorporated into the overall project. Such was the case in the design of an automated landing system for a major international airport (see bibliography).

MEDICAL TESTS

An interesting recent application of decision theory ideas was concerned with the screening of pregnant women for defects in the foetus. Two tests were available for determining whether a foetus suffered from spina bifida. The first was an easy-to-administer blood test, but this was not very accurate. Since a woman whose foetus was thought to have spina bifida would be offered an abortion, an inaccurate test could lead either to a healthy foetus being aborted or to an unhealthy foetus remaining undetected. A second test was available which involved drawing fluid from the womb. This test carried a much greater (although not 100%) accuracy, but involved a slight risk that the test itself might trigger a miscarriage. The problem with this type of test is, of course, that in screening for the less than 1% of foetuses with spina befida, a small percentage of the other 99%+ would miscarry.

A number of screening plans were drawn up based on the two available tests. For example, one screening plan might be:

Decision	whether to take a blood test.
Chance event	result of blood test.
Decision	in case of positive blood test result, whether to take a fluid test.
Chance event	result of fluid test.
Decision	in case of a positive fluid test result, whether to abort.
Chance event	outcome of pregnancy.

This is in the form suited to decision analysis which was used to compare the plan above with other screening plans and the option not to screen at all. The result was the decision to screen only women thought in advance to be at risk (older women and women with a family history of spina bifida).

Theory and practice of decision analysis

To demonstrate the technique, another simple example will be used. It is intended to illustrate decision analysis rather than be a practical decision problem.

Example

A company is in a position to market a new product. The decision must be made either (a) to go ahead and launch, or (b) to run a test market, or (c) to abandon the product. If the product is launched, market research shows that the probabilities of a strong, weak or non-existent market will be respectively, 40%, 40%, 20%. If a test market is used (at a cost of £30,000), it will give either favourable or unfavourable indications. The probability of a favourable indication has been calculated at 50%, of unfavourable also 50%. If the indication is favourable and the product is launched, the probabilities of strong, weak and non-existent markets are 72%, 24%, 4% respectively; if unfavourable the probabilities are 8%, 56%, 36%. The payoffs for strong, weak and non-existent markets are, respectively £200,000, £50,000, −£150,000. Expenses already incurred have been ignored since they are not relevant to the decision now faced. What decision should be taken? Table 8.1 summarizes the data.

Table 8.1

Demand	Payoff (£'000)	Probabilities		
		No test	Favourable test	Unfavourable test
Strong	200	0.40	0.72	0.08
Weak	50	0.40	0.24	0.56
Non-existent	−150	0.20	0.04	0.36

THE TREE

Method

Draw the decision tree. Think out the logical sequence and inter-relationships between decision nodes and chance event nodes. Usually one starts with a single decision node, followed by one or more chance nodes, then more decision nodes, then more chance nodes, and so on. The final decision tree consists of decision nodes, chance nodes and the straight lines showing their sequence. The pattern of decision nodes/chance nodes/decision nodes, etc. is typical but should not be followed slavishly, for there are circumstances where a series of decision nodes is followed by a further series of decision nodes and similarly for chance nodes.

Example

The first decision is launch, test market or abandon, and gives rise to the decision node at A. The decision to 'launch' leads to the chance node at B and 3 outcomes of a strong, weak or non-existent

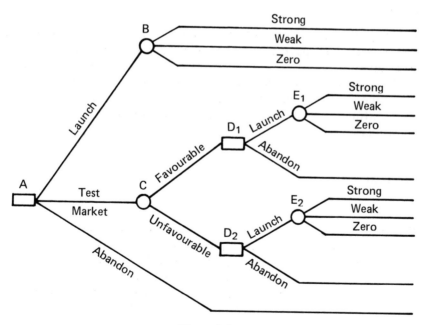

Figure 8.3

market. The decision to test market leads to the chance node at C indicating the result of the test market, either favourable or unfavourable. Faced with these new data one has then to decide whether to launch or abandon and the decision nodes at D1 and D2 show these choices. Having launched, the next stage is the outcome of the launch, a strong, weak or non-existent market, shown by the chance nodes at E1 and E2. At any stage a decision to 'abandon' ends that part of the decision tree. The resulting decision tree is shown in figure 8.3.

PAYOFFS

Method

Fill in the payoffs at the ends of all branches of the decision tree. The payoffs are usually but not necessarily in monetary terms; they should be carefully calculated so that irrelevant costs/revenues are excluded and so that inclusions/exclusions are consistent from branch to branch.

Example

The payoffs have been written into the decision tree as shown in figure 8.4. The payoff for the 'abandon' decision is zero, since costs already incurred have been excluded. The payoffs for strong, weak and non-existent markets are 200, 50, −150 as in table 8.1. Note that 30 has been subtracted from these amounts for the branch representing the 'test market' decision since this is the cost of running a test market.

PROBABILITIES

Method

The branches leaving chance nodes have assigned to each of them a probability indicating the likelihood of the specified event happening. The calculation of the probabilities may have come from market research, subjective assessment or any of the ways in which probabilities are determined. Branches leaving decision nodes do not have probabilities attached since the decision-maker determines along which branch to progress.

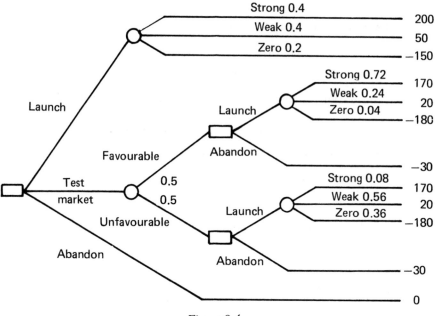

Figure 8.4

Example

The probabilities, taken from table 8.1, have been added to the relevant branches in figure 8.4. Note that the sum of the probabilities of the branches leaving each chance event node must sum to 1.0 since the branches at the node are supposed to cover all possibilities at that chance event.

ROLL-BACK PROCEDURE

Method

The analysis from this point consists of calculations with the payoffs and probabilities and is known as the *roll-back procedure* since it starts at the payoffs and works back to the initial decision.

(a) For each chance node the EMV is calculated.
(b) For each decision node the EMVs of the branches are compared and all but the most favourable branch are eliminated.

Example (see figure 8.5)

At B: EMV = (200 x 0.4) + (50 x 0.4) − (150 x 0.2)

$\qquad$ = 80 + 20 − 30

$\qquad$ = 70

It may be easier to understand this calculation by doing it in the same way as for the earlier example on coin tossing. Think of repeating the event 100 times and work out the average return.

At E1: EMV = (170 x 0.72) + (20 x 0.24) + (−180 x 0.04)

$\qquad$ = 122.4 + 4.8 − 7.2

$\qquad$ = 120

At E2: EMV = (170 x 0.08)) + (20 x 0.56) + (−180 x 0.36)

$\qquad$ = 13.6 + 11.2 − 64.8

$\qquad$ = −40

At D1: Eliminate the 'abandon' decision which has an EMV of −30 compared to the 'launch' EMV of 120.0. The elimination of a branch is denoted by ‖.

At D2: Eliminate the 'launch' decision which has an EMV of −40, compared to −30 for 'abandon'.

At C: $\quad$ EMV = (120 x 0.5) + (−30 x 0.5)

$\qquad$ = 60 -- 15

$\qquad$ = 45

At A: $\quad$ Eliminate the 'abandon' and 'test' branches which have EMVs of 45 and 0, compared with 70 for 'launch'.

ROUTE

Method

Finally, the path to be taken by the decision-maker through the problem should be summarized clearly.

Example

The decision-maker should go ahead and launch the product.

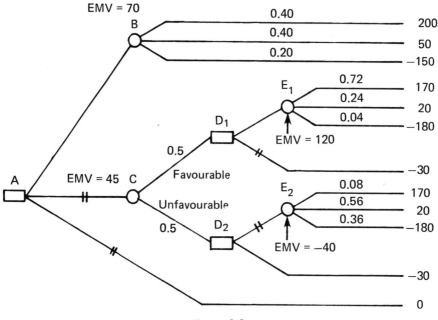

Figure 8.5

To summarize, a decision analysis problem can always be structured in terms of the five stages shown above:

(1) Draw the tree.
(2) Calculate the payoffs for the end of each branch.
(3) Insert probabilities for each branch stemming from a chance node.
(4) Roll-back.
(5) Define the decision route through the tree.

Additional information

Additional information can be obtained from the tree. In the product launch example, the EMV without the test market is 70.0. The EMV with the test market, but ignoring the cost of the test market is 75.0 (45.0 + the cost of the test market). The expected value of the test market information is therefore 5.0 (75.0 − 70.0). The amount is known as the *expected value of sample information* or EVSI. It is the average extra payoff that accrues with the benefit of the sample (or market) information. EVSI is a guide to the maximum amount that should be paid for a particular piece of sample information. If some market research were offered at a cost greater than its EVSI, it should not be taken up as its cost exceeds its benefit.

Suppose the test market were not carried out but, by some magical means, it were known in advance what the outcomes at all the chance nodes were going to be. How much profit could be made then? There are 3 possible outcomes: a strong, weak or non-existent market. If it were known which was going to occur, appropriate action would be taken.

Outcome	Action	Payoff
Strong	Launch	200
Weak	Launch	50
Non-existent	Abandon	0

With this perfect information the expected profit is:

$$200 \times P(\text{strong}) + 50 \times P(\text{weak}) + 0 \times P(\text{non-existent})$$

$$= (200 \times 0.4) + (50 \times 0.4) + 0$$

$$= \underline{100.0}$$

The expected value of knowing the outcome of chance events is:

Expected profit with this knowledge – Expected profit without any additional information

$$= 100.0 - 70.0$$

$$= 30.0$$

This amount is known as the *expected value of perfect information* or EVPI. When calculating EVPI it is tempting to believe that with perfect information the largest profit shown on the decision tree can always be made, e.g. in this case to suppose that 200 can always be made. This is incorrect since perfect information does not guarantee the best outcome; it merely tells what the outcome is, whether good or bad.

EVPI is the maximum amount that should be paid for any sample information. No survey or market research could give a benefit greater than the EVPI, so if its cost is greater than the EVPI it can be rejected immediately without bothering to incorporate it into the decision tree. In this example any market research, survey, etc., costing £30,000 or more would be immediately rejected.

To summarize, the EVSI refers to a particular piece of sample information and is the maximum to be paid for that *particular* information. EVPI refers to perfect information and is the maximum to be paid for *any* information.

Bayes theorem

In the product launching example it was assumed that the necessary probabilities were known or, at least, had been subjectively estimated. The relevant probabilities were:

$$\left.\begin{array}{l} P(S) = P(\text{strong demand}) = 0.4 \\ P(W) = P(\text{weak demand}) = 0.4 \\ P(N) = P(\text{non-existent demand}) = 0.2 \end{array}\right\} \ 1$$

$$\left.\begin{array}{l} P(S \,|\, F) = P(\text{strong, given a favourable test}) = 0.72 \\ P(W \,|\, F) = P(\text{weak, given a favourable test}) = 0.24 \\ P(N \,|\, F) = P(\text{non-existent, given a favourable test}) = 0.04 \\ P(S \,|\, U) = P(\text{strong, given an unfavourable test}) = 0.08 \\ P(W \,|\, U) = P(\text{weak, given an unfavourable test}) = 0.56 \\ P(N \,|\, U) = P(\text{non-existent, given an unfavourable test}) = 0.36 \end{array}\right\} \ 2$$

1 These are called *prior probabilities*. They are prior in the sense that they are the probabilities estimated prior to obtaining extra information from a test market.

2 These are called *posterior probabilities*. They are posterior in that they are probabilities of different demand levels after extra information has been obtained from a test market. Note the notation. The probability of a strong demand, given a favourable test, is written with a vertical line or stroke:

$$P(\text{strong} \,|\, \text{favourable}) \text{ or } P(S \,|\, F)$$

Both prior and posterior probabilities can be estimated subjectively using the correct assessment methods. Prior probabilities are nearly always estimated in this way. Posterior probabilities, on the other hand, are often estimated in a different way using:

(a) the prior probabilities; and
(b) information about the known accuracy of a particular type of test market.

The prior and posterior probabilities will then be consistent with one another; they may not have been had they been assessed independently. The method of calculating posterior probabilities from prior probabilities and the accuracy level of the test is called *Bayes theorem* (figure 8.6). Information about the known accuracy of the test is usually in the form: if there really is a strong market demand, then the probability that the test will give a favourable result is X, the probability that the test will give an unfavourable result is Y, etc. Putting this more mathematically, suppose information about the accuracy of the test in the product launching decision is:

$$P(F/S) = 0.9$$
$$P(U/S) = 0.1$$
$$P(F/W) = 0.3$$
$$P(U/W) = 0.7$$
$$P(F/N) = 0.1$$
$$P(U/N) = 0.9$$

This information has *not* been calculated. It was given by the devisers of the test market who used their knowledge of the structure of the test and their experience of using it to establish the above probabilities. The accuracy of tests (whether in marketing, medicine, etc.) is in this form because this is how the tests are checked; i.e. their performances in known situations are measured.

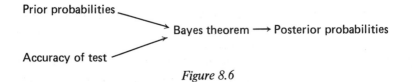

Figure 8.6

One can now show mathematically and diagrammatically what Bayes theorem does (see figure 8.7). Bayes theorem is a series of formulae for calculating the posterior probabilities P(S/F), etc. As an alternative to the algebraic method of calculating the posterior

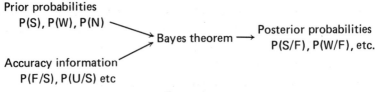

Figure 8.7

probabilities, using the formulae, a Venn diagram can be used. To do the calculations for the product launch example, the first step is to divide a rectangle into 3 parts representing strong, weak and non-existent demand levels as in figure 8.8.

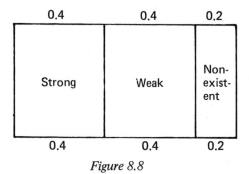

Figure 8.8

The second step is to subdivide each of these 3 parts into 2 parts to represent the outcome of the test market. The area referring to strong market is thus divided into an area which is nine-tenths of the whole and represents a favourable test market and an area which is one-tenth of the whole and represents an unfavourable test market (because $P(F/S) = 0.9$ and $P(U/S) = 0.1$). Similarly subdivisions are made for the weak and non-existent demand levels as in figure 8.9.

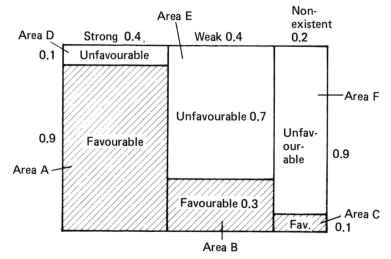

Figure 8.9

From figure 8.9, since probabilities are associated with areas, the required probabilities can be calculated. For example:

P(favourable test market)

   = total area associated with a favourable test

   = Area A + Area B + Area C

   = (0.9 x 0.4) + (0.3 x 0.4) + (0.1 x 0.2) = 0.5

P(strong demand given a favourable test market)

   = proportion of area associated with a favourable test which relates to a strong demand.

$$= \frac{\text{Area A}}{\text{Area A + Area B + Area C}}$$

$$= \frac{0.9 \times 0.4}{(0.9 \times 0.4) + (0.3 \times 0.4) + (0.1 \times 0.2)}$$

$$= \frac{0.36}{0.50}$$

   = 0.72

Similar calculations can be made for the other posterior probabilities. They are summarized in table 8.2. The posterior probabilities shown in the last column are the ones used when the problem was analysed earlier. As always, an intuitive check on the probability manipulations can be made by working in frequencies.

   In general, Bayes theorem is used whenever probabilities are revised in the light of some sample information. The resulting posterior probabilities are likely to be more accurate than if they had been estimated subjectively.

Validity of the EMV criterion

The criterion on which the decision has been based is that of EMV. The decision branch with the highest EMV was selected at each decision node. Many other criteria could have been used. For example, an alternative criterion is that of *maximin*, which means that the decision is chosen whose worst possible payoff is better than the worst possible payoff for any other decision (hence *maximin*).

Table 8.2 Bayesian analysis of product launch problem

Market test outcome F or U	Demand level in market S, W or N	Prior probabilities P(S), P(W), P(N)	Test accuracy P(F/S), etc.	Col. (3) × Col. (4)	Posterior probabilities P(S/F), etc. Col. (5) ÷ Sum Col. (5)
(1)	(2)	(3)	(4)	(5)	(6)
F	S	0.4	0.9	0.36 (Area A)	0.72
	W	0.4	0.3	0.12 (B)	0.24
	N	0.2	0.1	0.02 (C)	0.04
				P(F) = 0.50	
U	S	0.4	0.1	0.04 (D)	0.08
	W	0.4	0.7	0.28 (E)	0.56
	N	0.2	0.9	0.18 (F)	0.36
				P(U) = 0.50	

In the product launch example, maximin requires the abandon decision to be chosen, since for the launch decision the worst payoff is a loss of 150 and for 'test market' the worst payoff is a loss of 180, whereas for the abandon decision the worst is 0. Abandon has, therefore, the highest worst payoff.

EMV is the criterion most frequently used, but it suffers from one major drawback. The 'averaging-out' process of EMV sometimes masks a catastrophic payoff which has very low probability. Sometimes this outcome may be such that the decision-maker would not risk it however low the probability and however high the payoff if things go well. For instance, the decision to market might have had a fourth outcome which was a 1% chance of losing £1 million (in a legal case, perhaps). Although this would not alter the optimal EMV decision to 'market', if the loss of £1 million would bankrupt the company the decision-maker might prefer not to take the risk. He would, in other words, prefer the lower EMV decision of 'abandon' in order to avoid even a small probability of bankruptcy.

It is the same argument in reverse that persuades people to bet on football pools. The EMV of a £1 stake on football pools is negative since the total stake money finances, as well as the winning payouts, administrative expenses and no doubt a scandously small profit margin. The EMV decision would be not to place the bet. However, while the punter can probably afford to lose £1, the bet carries a small chance of winning an amount so large that it would completely and forever change the lifestyle of the punter (and there is virtually no other way, legal or illegal, of quickly amassing so large an amount).

A decision theory analysis should incorporate a check that the EMV has not in fact 'averaged out' some catastrophic but low probability payoff. The check is made by listing all the possible outcomes, under the assumption that the optimal decision strategy is followed. It is easy to do this in the product launch example (table 8.3). There may be more difficulty in other cases. The probabilities should sum to 1.0, otherwise an error has been made. The range of payoffs, with an 80% chance of making a profit, 20% of making a loss, is not unusual and seems acceptable. In other situations an analysis of the final outcomes either may highlight the possibility of a catastrophic payoff or may reveal that while the EMV appears healthy a loss is more likely than a profit. In both cases the decision to go ahead would depend upon the company's attitude to risk.

Table 8.3

Possible outcomes	Payoff	Probability
Strong	200	0.4
Weak	50	0.4
Non-existent	−150	0.2
		1.0

Worked examples

1. S. TRAIN SCHOOL OF MOTORING

A driving school renews its fleet of cars every 2 years. By replacing the entire fleet at one time favourable purchase prices are obtained. The size of the present fleet is 200. The managing director is aware that, depending upon the economic climate, the size of the next fleet may be more or less than 200. He estimates that if he buys 250 cars he could earn a profit of £250,000 in the next 2 years given an economic upswing, could earn a profit of £140,000 if things stay as they are, and could lose £50,000 if there is a recession. If he buys 200 cars (keeping the fleet at present strength), the earning will be respectively £180,000, £150,000 and £50,000. Buying only 150 cars would earn £150,000, £120,000 and £80,000 respectively. He estimates the chance of an economic upswing at 0.3, of staying level at 0.4 and of recession at 0.3. How many cars should the firm buy? The managing director is aware that he can purchase a survey by a well-known economic forecasting team which will give him a clear picture of economic prospects. What is the maximum he should pay for it?

Following the first four of five stages of a decision analysis gives figure 8.10. Forming the decision tree is straightforward, as is the attachment of the probabilities and payoffs which are given and do not require any calculation. The roll-back procedure gives EMVs as shown above each chance event node. For example, the chance event node A has an expected value of 116 calculated:

$$(250 \times 0.3) + (140 \times 0.4) + (-50 \times 0.3) = 116$$

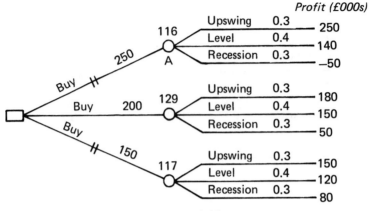

Figure 8.10

The middle branch of the tree has the highest EMV. The upper and lower branches are therefore eliminated. This leads to the fifth and final stage of decision analysis, summarizing the route. In this case it is simple: buy 200 cars.

The maximum amount the MD should pay for extra economic information is the expected value of perfect information. With perfect information action would be taken as shown in table 8.4.

$$\text{Expected value} = (250 \times 0.3) + (150 \times 0.4) + (80 \times 0.3)$$

$$= 75 + 60 + 24$$

$$= 159$$

Since a profit of 129 can be made without the information:

$$\text{EVPI} = 159 - 129$$

$$= 30$$

The maximum to pay for the extra information is £30,000

Table 8.4

If the information said the economic climate would be:	Probability	Action	Payoff
Upswing	0.3	Buy 250	250
Level	0.4	Buy 200	150
Recession	0.3	Buy 150	80

2. SOUTH EASTERN TELEVISION

This regional TV company has made a pilot programme of a newly devised situation comedy. The pilot progamme has been screened in its own region and was well received. The company now has the option of selling the idea and the pilot (for £75,000) to a rival company to develop, or it can decide to develop the pilot into a series itself. In the latter case it is thought to be equally likely that the series be successful or not. If the series were unsuccessful the idea would be abandoned, having lost the company £50,000. If the first series were successful another series would be made. Again, it is thought to be equally likely that the second series would succeed or fail. The failure of the second after a successful first would bring the company a total profit of £20,000. The failure of the second series would result in the programme being abandoned.

Table 8.5

Option	Payoff	Probability (%)
Major success	£3 million	20
Minor success	£0.5 million	50
Flop	−£1.5 million	30

Payoffs include profits on the first 2 series.

Two successful series would bring the option of making a feature film for the cinema circuit. Three possibilities could arise should a feature film be made. The probabilities and payoffs are shown in table 8.5. Merely continuing the programme for further TV series could be expected to make a total profit of £350,000 including the profits from the first 2 series.

(a) What should the company do?
(b) If the optimal decision is followed, what is the probability that the company would make a loss?

The decision tree is drawn, the payoffs and probabilities inserted resulting in figure 8.11. It illustrates the fact that it is possible to have consecutive chance nodes (or consecutive decision nodes). The

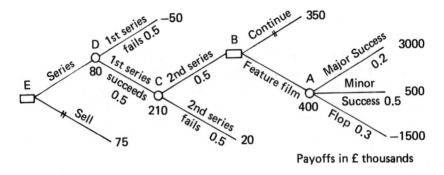

Payoffs in £ thousands

Figure 8.11

roll-back procedure starts at node A where the EMV is:

Node A   EMV = (0.2 x 3000) + (0.5 x 500) + (0.3 x −1500)

= 600 + 250 − 450

= 400

Node B   Rolling back to the decision node B, the choice is between a decision with an EMV of 350 and one with an EMV of 400. The 350 branch (continue with TV series only) is eliminated.

Node C   EMV = (0.5 x 400) + (0.5 x 20)

= 200 + 10

= 210

Node D   EMV = (0.5 x −50) + (0.5 x 210)

= −25 + 105

= 80

Node E   The decision to make the series is preferred to the sell option since its EMV is 80 as against 75.

The route through the tree is:

(1) Make the pilot programme into a series
(2) If two series succeed, take up the option to make a feature film.

Following this optimal route the possible final outcomes are shown in table 8.6. The overall probabilities of these outcomes are calculated

Table 8.6

Outcome	Payoff	Probability	
1st series fails	−50	0.5	
2nd series fails	20	0.25	(= 0.5 x 0.5)
Film is major success	3000	0.05	(= 0.5 x 0.5 x 0.2)
Film is minor success	500	0.125	(= 0.5 x 0.5 x 0.5)
Film is flop	−1500	0.075	(= 0.5 x 0.5 x 0.3)
		1.000	

by multiplying the probabilities of the individual events which make up the final outcome. Alternatively, the outcomes can be viewed in frequency terms (as with expected value). If the company embarked on this decision 100 times, the 1st series could be expected to fail 50 times. Of these 50, the 2nd series could be expected to fail 25 times (probability = ½). The probability of the outcome '2nd series fails' is therefore 25/100 = 0.25.

Note that the payoff of £350,000 is not possible since that decision branch has been struck off.

Two of the 5 outcomes involve a loss. Their probabilities are 0.5 and 0.075. The probability of a loss is therefore 57.5%.

In spite of the healthy EMV of £80,000, the company is more likely to make a loss than a profit. Further, one of the possible losses is a very large one (£1.5 million). In view of this, and the fact that the alternative to the EMV of £80,000 is both safe (100% probability) and close to £80,000 (£75,000), the company may wish to ignore the EMV recommendation and sell the pilot programme. This is a further illustration that the EMV optimal decision is a guide only.

3. BELFAST AND IRELAND BREWERIES

The marketing manager of the Belfast and Ireland Breweries is concerned with the sales appeal of the firm's present label for its half-litre canned beer. Market research studies indicate that supermarket customers find little 'eye-appeal' in the labels. The firm's design artist has produced some prototype labels in different styles, all of which have been evaluated by the firm's executives. One design has consistently won in all the preference tests. However, the marketing manager is still in some doubt as to whether the new labels would increase sales appreciably. He considers the costs associated with converting machinery, inventory and point-of-purchase displays, etc. and concludes that an out-of-pocket cost of £50,000 would be involved.

If the new label were really superior to the old, the marketing manager estimates that the present value of all net cash flows related to the increased sales and costs generated over the next 3 years by the more attractive label would exceed by £80,000 the cash flows anticipated under the present label. Based on his prior experience with merchandising changes of this type, he is only willing to assign a 50–50 chance to the event 'new label superior to old'.

(a) What course of action maximizes expected monetary value (EMV)?

(b) Suppose it is possible to delay making a decision in order to gather more information about the likelihood of the new label being superior. Specifically, suppose the marketing manager can purchase for £10,000 a survey which is claimed to have a 90/80 reliability. That is, if the new label is superior, there is a 90% chance that the survey would confirm this superiority; if the new label is not superior, there is an 80% chance that the survey would confirm this fact. What action should the manager take?

(a) The decision tree is a simple one (figure 8.12). The optimal decision is to introduce the new label. It has an EMV of £15,000. Note that the assumption is being made that if the new label is not superior, it is not inferior either. The worst that can happen is that the costs of the conversion would be lost.

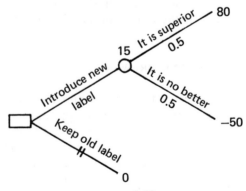

Figure 8.12

(b) If the survey is considered, an extra branch should be added to the decision tree as shown in figure 8.13. Note the way in which the cost of the survey is handled in figure 8.13. A 'gate' ⸭ is drawn at the beginning of the branch. This saves the trouble of deducting the £10,000 from each payoff. The result will be the same in each case. The probabilities in figure 8.13 are calculated from Bayes theorem. The prior probabilities were:

$$P(\text{superior}) = 0.5$$

$$P(\text{not superior}) = 0.5$$

They are revised in the light of information from the survey. The posterior probabilities are of the form:

$$P(\text{superior/survey says superior})$$

$$P(\text{superior/survey says not superior}), \text{etc.}$$

Construct a Venn diagram as in figure 8.14. First divide the square according to the prior probabilities. Further sub-divide the square according to the known reliability of the survey. If the survey has said that the new label is superior, areas A and C only are relevant.

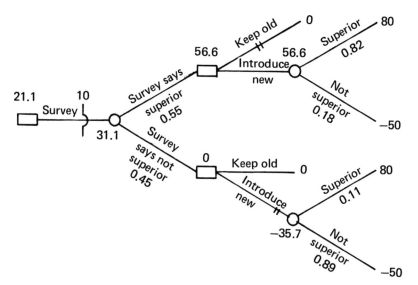

Figure 8.13

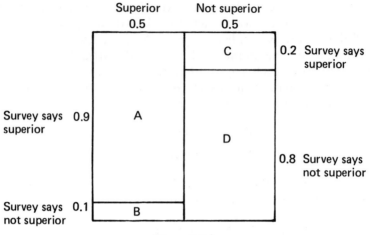

Figure 8.14

Within this area, A refers to the occasions when the new label actually is superior. Thus:

$$P(\text{superior/survey says superior}) = \frac{A}{A + C}$$

$$= \frac{0.45}{0.55}$$

$$= 0.82$$

Similarly:

$$P(\text{superior/survey says not superior}) = \frac{B}{B + D}$$

$$= \frac{0.05}{0.45}$$

$$= 0.11$$

$$P(\text{not superior/survey says superior}) = \frac{C}{A + C}$$

$$= \frac{0.10}{0.55}$$

$$= 0.18$$

$$P(\text{not superior/survey says not superior}) = \frac{D}{B + D}$$

$$= \frac{0.4}{0.45}$$

$$= 0.89$$

$P(\text{survey says superior}) = A + C = 0.55$

$P(\text{survey says not superior}) = B + D = 0.45$

The overall EMV associated with the 'survey' branch works out to be £21,100 after following through the roll-back procedure. This is higher than the EMV (£15,000) when there was no survey. The optimal strategy is:

(1) Carry out a survey.
(2) (a) If the survey says 'superior' then introduce the new label.
    (b) If survey says 'not superior' then keep the old label.

The range of possible outcomes under the optimal strategy is shown in table 8.7. There are no catastrophic outcomes, but it should be noted that there is a higher probability of making a loss (55%) than a profit (45%).

Table 8.7

	Payoff	*Probability*	
Old label kept	−10	0.45	
New label superior	70	0.45	(= 0.82 x 0.55)
New label not superior	−60	0.10	(= 0.18 x 0.55)
		1.00	

Final comments

Decision analysis (and most other techniques) should be regarded as an aid to the decision-making process, not a substitute for it. Having applied decision analysis its recommendations should be tested against other criteria – is the decision tree an accurate model of the decisions faced? How much faith can be put in the probability estimates? Do the reservations about the EMV criterion apply in this case? Only if the assumptions that have permitted the use of decision analysis are realistic can the recommendations be accepted at face value; otherwise the recommendations should be regarded as a useful guide to the final decision. It may be advisable to carry out some sensitivity analysis by repeating the procedure, each time varying those assumptions which are the most doubtful. This will demonstrate whether the optimal decision would be different if the assumptions were different.

Sensitivity analysis may involve the probabilities of the chance events, since they are sometimes regarded with suspicion by the decision-makers. This is especially true since decision analysis often requires the subjective assessment of probabilities. This must be done well and painstakingly if the end-result is to be sensible. Methods have now been developed of making the assessments as accurately as possible, and they should always be used in decision analysis. Nevertheless, it is the idea of subjective probabilities that causes most difficulties with the technique and dissuades many decision-makers from using it.

Decision analysis can be extended to include continuous distributions as the outcome of a chance event. In the examples of this chapter only discrete distributions have been used. Only a small amount of additional theory is needed to incorporate continuous ones. For example, the outcome of a product launch, instead of being the discrete payoffs £200,000, £50,000, −£150,000, could have been specified by a continuous distribution.

Even if the recommendations of the technique were rejected by the decision-maker, use of the technique will still have fulfilled an important objective. Decision analysis requires the structuring of the decision and the true sequence and logical relationships between 'sub-decisions' and chance events to be understood. However the final decision is arrived at, it should now be based upon a deeper understanding of the nature of the problem.

Further examples

1. NATIONAL UNITED TRADES

The purchasing manager of National has to decide what to do about forward dealings in oranges. Two cargoes are to be landed in London in January next year, one of 500 tons and one of 400 tons. The purchasing manager can buy either at £240 per ton, but he cannot afford to buy both. He must sign a contract of purchase now or lose the chance to buy. Smaller quantities of up to 200 tons can usually be purchased at any time for immediate delivery, at a price of £270 per ton.

There are two main customers for oranges, fruit and vegetable wholesaling chains: Finefruit (FF) and Home Produce (HP). In January FF require 500 tons and HP 350 tons. The market price for such quantities is likely to be £320 per ton. They will each place a contract with a single supplier, and National's reputation is such that they believe they have a good chance of winning either contract. Competition is usually on past record, rather than price. If National should fail to win either, it can sell small quantities (up to 50 tons) at £290 per ton, but larger quantities would have to be sold at clearance prices of £240 per ton. National is unable to store oranges.

National's sales manager has been debating his chances of winning either FF's or HP's contracts. He cannot bid for both because the two companies insist on more than one supplier. If National purchases the 500-ton cargo, the sales manager feels he has a 50 : 50 chance of winning the FF contract. Since the wholesalers seek above all security of supply, the sales manager feels that if he buys the large cargo he has an 80% chance of winning the HP contract but only a 70% chance if he buys the small cargo. It would be possible to buy the small cargo and then bid for FF's contract relying on small purchases to make up the difference, but since FF would be uneasy about his ability to supply, he feels his choice of winning the contract would be only 40%.

What is National's optimal strategy?

2. CO-ORDINATION (HOME AND OVERSEAS) SERVICES

The company acts as a co-ordinator on electrical projects. A client specifies a project and then hands it over to Coordination Services who try to put together a team of subcontractors who can complete the project according to the specified standards and price. The company rarely do any of the work themselves.

The advantage to the client is that they deal only with one company instead of a range of companies. Nor does the client have to deal with the technicalities of the project. In return a commission is received which is a percentage of the total value of the project. If the company cannot organize a team to meet the client's specifications then they receive no payment.

Recently the company has been trying to expand its overseas business. It now has an opportunity to work for an African government, that has specified three projects (table 8.8). There is a commission of 10% on any of the projects co-ordinated. The government, however, are wary of dealing with western companies and have laid down the following conditions:

'Project A is to be arranged first. If Co-ordination Services cannot assemble a satisfactory team within 6 months, then they must drop out without having a chance to co-ordinate projects B and C. On the other hand if they are successful with A, they will collect their commission and have the option to stop or to try to co-ordinate a second project under the same conditions (i.e. assemble a team within 6 months or no commission and no chance to co-ordinate the third project). If they meet with success on the first two projects they have the option to try to co-ordinate the third.'

This was an unusual deal and Co-ordination Services' board was also wary of the government. The marketing director proceeded to analyse the proposal. He estimated the costs of trying to co-ordinate the projects and the chances of successful outcomes given the specifications as shown in table 8.9. Since the projects were so different in the nature of the expertise they entailed, it was thought that successfully arranging one project would not make the successful arranging of the other two any more or less likely. The costs would be incurred whether or not teams were successfully assembled, but of course no costs would be incurred with respect to a project if the company did not attempt to co-ordinate it.

Since project A would have to be co-ordinated successfully before either B or C could be attempted, the marketing director prepared a

Table 8.8

Project	Value
A. Aircraft tracking system	£250,000
B. Buildings security	£800,000
C. Conservation of fuel monitoring devices	£2,000,000

Table 8.9

		Probabilities	
Project	Cost	Success	Failure
A	£20,000	0.6	0.4
B	£10,000	0.8	0.2
C	£15,000	0.5	0.5

table showing profits/losses that could be anticipated for project A (table 8.10). Thus, based on expected value project A would be unprofitable. However, success with project A would return more

Table 8.10

Event	Probability	Payoff
Project A successfully co-ordinated	0.6	£5,000
Project A not co-ordinated	0.4	−£20,000
Expected value = 0.6 x £5000 − 0.4 x £20,000 = −£5,000		

than just commission. It would also give the opportunity to attempt either B or C, which both have lower associated costs and higher commissions.

What recommendation would you make to the board?

What is the expected value of perfect information to the company?

3. SOUTH HAMPSHIRE ASSOCIATED PLASTIC ENTERPRISES

The R & D manager for Plastic Enterprises is trying to decide whether or not to fund a project to develop a new moulding process. In terms of the uses to which the process can be put the project could be a major success with an estimated payoff of £240,000. If it had a narrower range of uses it would be a minor success with an estimated payoff of £20,000. If the project fails completely it will cost the company £100,000.

The scientists involved and the manager assess probabilities as follows:

$$P(\text{major success}) = 0.10$$
$$P(\text{minor success}) = 0.50$$
$$P(\text{failure}) \quad\;\; = 0.40$$

Should the project be funded? Suppose a research institute offers to make a preliminary study, costing £35,000, and make a recommendation as to the likely outcome of the project. Should the offer be considered (i.e. is £35,000 less than the EVPI)?

As an alternative to the outside group, an experiment can be conducted internally to develop on small-scale prototype of the moulding process. It will be tested on four production materials. There are three possible outcomes to this experiment:

O_1 : the prototype works well on all four materials;
O_2 : the prototype works well on three of the materials;
O_3 : the prototype works well on fewer than three of the materials.

Based on their experiences turning other experimental processes into production operations, the conditional probabilities in table 8.11 were estimated.

Table 8.11

Probability of	Given		
	Major success	*Minor success*	*Failure*
O_1	0.75	0.30	0.10
O_2	0.20	0.60	0.40
O_3	0.05	0.10	0.50

If the experiment is conducted and O_2 is the outcome, should be project be funded?

What should the R & D manager's decision strategy be?

What is the expected value of sample information for the experiment?

Assuming the experiment is conducted, what is the probability that the project will be profitable?

ANSWERS

1. National United Traders

Buy the large cargo and bid for HP's contract. The EMV of this strategy is £24,900.

2. Co-ordination (Home and Overseas) Services

Recommendations: Attempt project A
                            If successful, attempt B
                            If successful, attempt C.

EMV = £68,200.
Expected value of perfect information = £23,900.

3. South Hants Associated Plastic Enterprises

(a) Do not fund. EMV = −£6000.
(b) Refuse. EVPI = £34,000.
(c) No. EMV = −£10,700.
(d) If cost of internal experiment is less than £17,000 then conduct it, otherwise abandon it. If experiment is conducted, then fund the project if the experiment outcome is 0_1, abandon it otherwise.
(e) EVSI = £17,000.
(f) Probability of profitability = 22.5%.

Linear programming

At the end of the chapter the reader should know sufficient about linear programming to formulate a problem, to solve graphically a small LP and to interpret the basic output provided by computer packages. The reader should be aware of the practicalities of applying LP.

Linear programming (LP) is a technique for solving certain types of decision problems. It is applied to problems which require the optimization of some criterion subject to a series of limitations restricting the actions that can be taken. It has widespread use in a variety of organizations, especially large ones. Its value is not confined to its problem-solving power, since its use encourages a clearer structuring and therefore understanding of the problem. Indeed, as with many similar techniques, the process of arriving at a solution often provides more benefit than the solution itself. Besides the actual solution, the technique provides a range of extra information about the problem.

Any decision problem necessarily has two elements. Firstly, there are decision variables, the setting of which constitutes the decision; secondly, there is an objective which the decision-maker hopes to achieve, or nearly achieve, through taking the best possible decision, i.e. the objective is a criterion which distinguishes a good decision from a bad decision. The objective might be the maximization of financial contribution and the decision variables the production level for each of the company's products. The decision-maker hopes to select production levels which will make the financial contribution as high as possible. In slightly more complicated, and realistic, problems, the decision variables might be, in addition, subject to some other relationships. For instance, the availability of some scarce resource required in the production process might limit the output range of the product. Such relationships are known as constraints. If the objective and the constraints can be expressed mathematically in terms of the decision variables, and if these mathematical expressions are linear (they contain no logarithmic, squared, cubed, etc., terms) then the problem can potentially be solved by LP.

At this stage it is important to grasp the overall structure of an LP problem. To reiterate, there are three parts to an LP problem:

- decision variables;
- objective function;
- constraints.

The objective function and the constraints are mathematically linear.

Formulating an LP problem

The first task in an LP problem is formulation. Given a verbal statement of the problem, the three elements (decision variables, objective function and constraints) have to be isolated and then the problem expressed algebraically.

EXAMPLE

This example has been simplified in order to illustrate the concepts underlying LP. A company makes two types of electronic circuits, type X and type Y, from two resources, machinery and manpower. The profits (in £) accruing from selling the products are shown in Table 9.1. The input of resources required to make the products are shown in Table 9.2. In other words, to produce 1 unit of product X, 5 machine-hours and 3 man-hours are used up; to produce 1 unit of product Y requires 2 machine-hours and 4 man-hours.

Table 9.1

	Product	
	X	Y
Profit/unit	4	3

Table 9.2

	Product	
	X	Y
Machine-hours	5	2
Man-hours	3	4

The availability of machinery is 80 hours/day and that of man-power is 90 hours/day. The decision problem is to determine the quantities of X and Y to produce per day in order to maximize profit.

To formulate the problem as a linear program, look, in turn, at the three basic elements.

(a) *Decision variables*

The decision to be made is the amounts of X and Y to produce. These quantities are then the decision variables:

x = the production level of product X;
y = the production level of product Y.

(b) *Objective function*

The aim is to make as much profit as possible. The function which expresses profit in terms of the decision variables must be determined. If 10 units of X are made and 20 units of Y, the profit would be:

$$10 \times 4 \ + \ 20 \times 3 = 100$$

Product   Product
X   Y

When the production levels are the values of the unknown decision variables x and y, the profit is:

$$4x \quad + \quad 3y$$

Product   Product
X Y

Consequently, the objective function is:

Maximize $4x + 3y$.

If the production levels were known, then substituting these values for x and y in the above would give the amount of profit made.

(c) *Constraints*

The amount of profit that can be made is limited by the availability of resources. Two constraints (one each for machinery and manpower) must be determined expressing the limits in terms of the decision variables. Suppose 10 units of X are made, then the resources are used up as follows (see table 9.2):

(10×5) machine-hours

(10×3) man-hours

Suppose 20 units of Y are made, then the resources used are:

(20×2) machine-hours

(20×4) man-hours

At these production levels, the *total* resource usage is:

$(10 \times 5) + (20 \times 2)$ machine-hours

$(10 \times 3) + (20 \times 4)$ man-hours

With unknown production levels x and y, the resource usages are:

$5x + 2y$ machine-hours

$3x + 4y$ man-hours

The availability of the resources is 80 and 90 respectively, thus the constraints may be written:

$5x + 2y \leqslant 80$ for machinery

$3x + 4y \leqslant 90$ for manpower

($\leqslant$ is mathematical shorthand for 'less than or equal to'). These constraints are inequalities, implying that the resources do not have to be used up. There may be spare capacity at the chosen production levels. The full linear programming formulation is:

$$\text{Max.} \quad 4x + 3y$$

$$\text{subject to} \quad 5x + 2y \leqslant 80$$

$$3x + 4y \leqslant 90$$

$$x, y \geqslant 0$$

($\geqslant$ is mathematical shorthand for 'greater than or equal to').

Both the decision variables must be restricted to be non-negative, since negative production levels would not be sensible. It is for this reason that the final two constraints, $x \geqslant 0$ and $y \geqslant 0$, are included.

Applications of LP

LP is applied to many business situations. Some organizations, for instance the oil companies, are major users of the technique. It is also true that LP is better suited to some problems than others. Here are some examples of its use with indications of why it is successful in each area.

CONTROLLING THE OPERATION OF AN OIL REFINERY

An oil refinery receives crude oil from tankers and converts it into a range of products such as motor car petroleum, diesel fuel, aircraft paraffins, etc. The crude oil varies in quality and type depending mainly on where it has come from. The profitability of the products changes in response to price changes and stock levels. LP is used to control the conversion process so that the most profitable mix of products is obtained.

The decision variables are the quantities of each product to make in a given production period; the objective function is the financial contribution which is to be maximized; the constraints are the amount of crude available, its quality and characteristics, the capacities of the refinery processes, available storage for products, etc.

Oil refining is a capital-intensive, continuous-production process which, relatively, is under the direct and immediate influence of the management. Changes in output suggested by the LP can be made fairly readily. It is for that reason that LP can be used to control the production process. Contrast this with the next application, where external factors mean that management cannot exercise an immediate influence and LP is used for planning not control.

SHORT-TERM PLANNING FOR A DISTRIBUTION SYSTEM

A manufacturing company has production plants at locations throughout the country. The output of each plant is transported, by road or rail, to one or more of the storage warehouses which are at (mostly) different locations. From the warehouses the products are delivered to customers. The company must decide how to distribute its products from the plants to the warehouses. Each month it plans how much production will be moved from each plant to each warehouse. It hopes to do this so that requirements at the warehouses are met in the cheapest possible way.

The decision variables are the quantities to be sent along each of the possible routes between plants and warehouses; the objective function is the cost of the distribution which has to be minimized. (Note that LP is used for maximization and minimization.) The constraints are the demands to be met at warehouses, the production available at plants and the transport capacities along each of the routes.

The distribution is not under such immediate and direct influence as in the oil refining case. Lorries break down, drivers are ill and so on. Management are not able to make instantaneously any changes suggested by the LP. For this reason LP is used to plan, say, the month ahead rather than for day-to-day control.

THE INGREDIENTS OF A PET FOOD PRODUCT

One of the earliest examples of the use of LP was in deciding the constituency of a brand of pet food. The product was to be made up from various ingredients: sago flour, meat meal, skimmed milk and others. The product has certain minimum requirements in terms of the proportions of protein and fat it is to contain. It has maximum requirements of the amounts of other factors such as fibre and ash. The ingredients each have different proportions of protein, fat, fibre, ash. etc. LP can be used to decide how much of each ingredient should be used so that the requirements are all met but the cost is kept to a minimum.

The decision variables are the percentages of each ingredient (sago, meat meal, etc.) the pet food is to contain. The objective function is the cost of a given amount (say, 1 ton) of the pet food. The constraints are the minimum and maximum requirements for the proportions of protein, fat, etc., contained in the product.

An example of the formulation of this type of problem is given at the end of the chapter. In a similar way LP can be used to decide on the ingredients of products other than pet foods.

Graphical solution

Dantzig's development of an algorithm (meaning a method of solution) for LP problems in the late 1940s combined with the increasing power of computers allows even large LP problems to be solved. Dantzig's algorithm is known as the Simplex method. Simplex, or variations of it, are used to solve even the largest-scale problems (with thousands of decision variables, thousands of constraints).

In practice computer packages are the means by which LP problems are solved. However, certain smaller problems can be solved graphically. This method gives insights into LP which help to interpret and understand the output and economic information provided by computer solutions. To gain these insights the simple example of electronic circuits formulated earlier will be solved by the graphical method. The formulation was:

$$\text{Max} \quad 4x + 3y$$
$$5x + 2y \leqslant 80$$
$$3x + 4y \leqslant 90$$
$$x, y \quad \geqslant 0$$

STAGE 1

The search is for a value of x and a value for y which satisfy the constraints and simultaneously make the objective function take on the highest value possible within the constraints. The first step is to represent the constraints graphically and thereby show the range of values of x and y which satisfy the constraints. The non-negativity constraints ($x \geqslant 0$, $y \geqslant 0$) restrict the search to the shaded area shown in figure 9.1, the whole of one quadrant. Only in this area are both x and y non-negative.

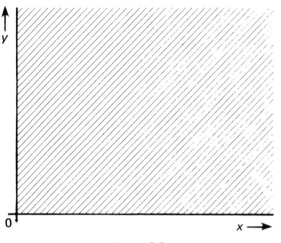

Figure 9.1

STAGE 2

For the first resource constraint ($5x + 2y \leqslant 80$) the line $5x + 2y = 80$ is drawn by the following method:

(a) Put $x = 0$. When $x = 0$, $2y = 80$, $y = 40$. The line therefore goes through the point ($x = 0, y = 40$). The line crosses the y axis at 40.
(b) Put $y = 0$. When $y = 0$, $5x = 80$ or $x = 16$. The line therefore crosses the x axis at 16.
(c) Join the two points on the axes to give the line $5x + 2y = 80$.

To satisfy the constraint, any pair of x, y values (in other words, any point) must fall *on or below* this line. Together with non-negativity constraints, the restriction is now to the shaded area shown in figure 9.2.

STAGE 3

For the second resource constraint ($3x + 4y \leqslant 90$), draw the line $3x + 4y = 90$. This is done, as before, by finding the points at which the line crosses the two axes. For this line the points are ($x = 0$,

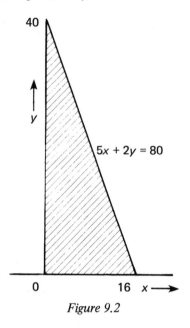

Figure 9.2

$y = 22.5$) and ($y = 0$, $x = 30$). To satisfy this constraint, any point must fall on or below this line. Together with the previous constraints, this means that the search for the optimal point is restricted to the shaded area shown in figure 9.3. This area is called the *feasible region*. Any point (pair of x, y values) within the feasible region satisfies all the constraints. The task is now to find which point in this area gives the objective function its largest value. Recall that the objective function is $4x + 3y$.

STAGE 4

Draw the line $4x + 3y = 30$ as shown in figure 9.4. All points on this line give the objective function the same value of 30 (the value of 30 was chosen arbitrarily). Since the line passes through the feasible region, there are many feasible points giving an objective function value of 30. Non-mathematically, this means that there are many combinations of production levels for the products which do not conflict with the availability of resources and which will result in a profit of 30.

Choose, again at random, another objective function value, say,

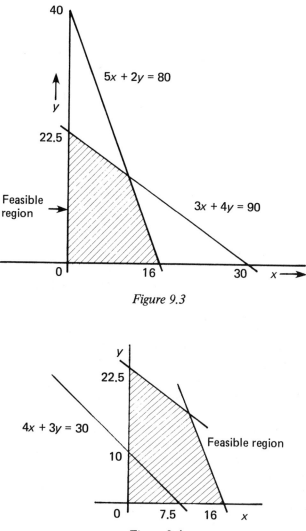

Figure 9.3

Figure 9.4

100. In figure 9.5, the line $4x + 3y = 100$ has been drawn. All points on this line give the same objective function value of 100. However, since this line does not pass through the feasible region, there is no point which satisfies the constraints and gives an objective function value of 100. A profit of 100 is thus impossible given the availability of resources.

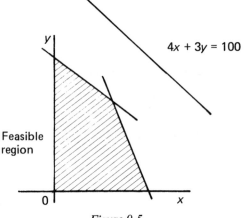

Figure 9.5

Return to the isoprofit line $4x + 3y = 30$. (Any line whose points give the same objective function value is called an *isoprofit line*). If it were to be moved, as in figure 9.6, parallel to itself in a north-easterly direction ($4x + 3y = 30 \Rightarrow 4x + 3y = 32 \Rightarrow 4x + 3y = 34 \Rightarrow$, etc.) the objective function value would gradually be increased. Indeed, if it were moved sufficiently far enough, the line would eventually coincide with $4x + 3y = 100$. However, the line is to be moved only far enough so that it still passes through the feasible region. This line will give the largest objective function value consistent with the constraints; the point (or points) that has remained within the feasible region will give the pair of x, y values that satisfy the constraints *and* maximize the objective function. In other words, this point (or points) is the solution to the LP problem.

Looking at figure 9.6, it can be seen that line L is the furthest an isoprofit line can be moved yet still intersect the feasible region. Point A is the only point on this line L that is within the feasible region. Any other isoprofit line either has a lower objective function value, or has no points within the feasible region.

The optimal point is A (the intersection of $5x + 2y = 80$ and $3x + 4y = 90$). The x and y values of A can be found by reading off from the graph or by simple algebra. The algebraic approach is as follows:

The point of intersection of two lines is to be found, i.e. x and y values that satisfy simultaneously two equations

$$5x + 2y = 80$$

$$3x + 4y = 90.$$

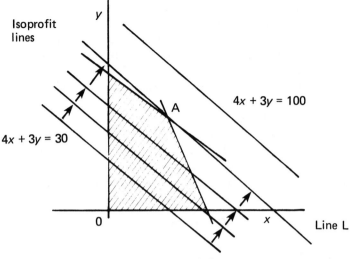

Figure 9.6

Multiply the first equation by 2. Leave the second equation alone and then:

$$10x + 4y = 160$$
$$3x + 4y = 90$$

Subtract the second equation from the first:

$$10x + 4y = 160$$
$$\underline{3x + 4y = 90}$$
$$7x = 70$$
$$\underline{x = 10}$$

Put $x = 10$ into one of the original equations

$$5x + 2y = 80$$
$$5 \times 10 + 2y = 80$$
$$2y = 30$$
$$\underline{y = 15}$$

The optimal point A is therefore $\underline{(x = 10, y = 15)}$

This algebraic method (known as simultaneous equations) is based on the general principle of multiplying one or both the original equations by constants so that when the resulting two equations are added or subtracted, one of the variables is eliminated. In the above example one of the equations was multiplied by 2. This meant that when the other was subtracted from it y disappeared, leaving an equation involving x only.

Having obtained the optimal point, the x and y values are substituted into the objective function to give the optimal value. Putting $x = 10$, $y = 15$ into $4x + 3y$ gives $4 \times 10 + 3 \times 15 = 85$.

The solution to the example is to produce 10 of product X, 15 of product Y. This production mix will give a profit of 85.

Computer solution

The graphical method is not a practical method of solution. In any case it could not be employed for a problem with more than two variables. A computer package would normally print out the optimal values of variables and objective functions without any need for graphs or simultaneous equations. The graphical method has been used here to give insights into LP. The way a computer solution works can be seen by consideration of certain aspects of the graphical method.

Firstly the point A is not always the optimal point. If, in the example, the objective function had not been $4x + 3y$ but had been $40x + 3y$, then the graphical solution would have looked as in figure 9.7. The optimal point is B ($x = 16$, $y = 0$) and the optimal objective function value is 640.

One of the corner points of the feasible region, however, is always an optimal point. As an objective function value increases and the line moves north-easterly out of the feasible region, it must always be a corner point that is the last point of contact between objective line and feasible region. The exception to this occurs when the objective line is parallel to a constraint. Then two corner points and the points on a line between them are all optimal points.

The Simplex method on which computer solutions are based makes use of the fact that one corner point is always an optimal point. The method starts with a solution at one corner of the feasible region. It then moves to a new solution at an adjacent corner, selecting that corner that brings about the biggest increase in the objective function. The process continues until no further movement brings about an increase in the objective. The optimal point has then been found.

The Simplex method quickly solves an LP because it does not

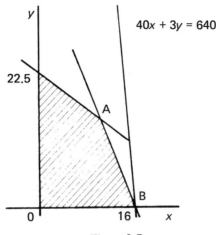

Figure 9.7

consider the whole of the feasible region, but only a small number of the corner points. Bear in mind that with a large LP the number of constraints and thus corner points is large, and therefore the Simplex method represents a considerable saving even over the alternative of evaluating the objective function at all corner points.

An LP problem may not always have a solution. Two situations arise in which a computer package will not provide a solution. The first situation is when the problem is *infeasible*. This means that it is impossible to satisfy all the constraints simultaneously and therefore no feasible region exists.

Suppose the previous example involving the products X and Y had an extra constraint, which was that production of X had to satisfy orders already received. If there were such orders for a total of 20 of product X, then the additional constraint would be:

$$x \geqslant 20$$

i.e. the minimum value x can take is 20.

The problem now becomes

$$\text{Max.} \quad 4x + 3y$$
$$\text{Subject to} \quad 5x + 2y \leqslant 80$$
$$3x + 4y \leqslant 90$$
$$x \geqslant 20$$
$$y \geqslant 0$$

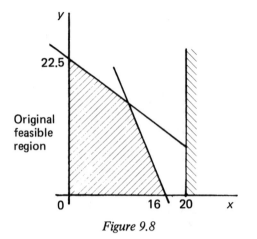

Figure 9.8

A graphical representation of this is shown in figure 9.8. No point within the original feasible region satisfies the new constraint. The problem is infeasible.

The second situation which does not produce an answer to a problem is when the solution is *unbounded*. Suppose in the above problem there were a minimum order requirement for Y of 12. Then an extra constraint would be added:

$$y \geqslant 12$$

Suppose also that the constraints on the resources were omitted, either because there were no limitations on these resources or, in error, they were forgotten. The problem would then be:

$$\text{Max.} \quad 4x + 3y$$
$$\text{Subject to} \quad x \geqslant 20$$
$$y \geqslant 12$$

Representing the problem graphically figure 9.9 results. The feasible region stretches to infinite values of x and y. Unlimited increases in x and y are possible, thereby improving the objective function yet all the while satisfying the constraints. The solution is *unbounded*.

Given the usual, but unfortunately not negligible, provisos that the data have been put into the computer correctly and that the computer is working, the output of the package will either be a finite solution or it will report that the problem is either infeasible or unbounded.

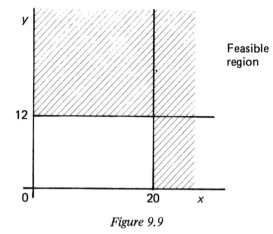

Figure 9.9

Dual values

When an LP problem is solved by a routine such as Simplex, besides providing the optimum values of the decision variables and the objective function, other information is given which can be used to perform sensitivity analysis. This means testing how much the optimal solution is affected if the basic assumptions or input data of the problem were changed. Included in this information are dual values, sometimes called shadow prices.

Dual values refer to constraints. Each constraint has associated with it a dual value. This is a measure of the increase in the value of the optimal objective function if 1 extra unit of the resource associated with that constraint were available, everything else being unchanged.

In the example, the dual values are as shown in table 9.3. The interpretation of the dual values is that if the available machine hours were 81 instead of 80, the objection function value could be increased by 0.5 to 85.5. So were the decision-maker offered an extra machine-hour at the same price as the previous 80 (the price already included

Table 9.3

Constraint	Dual value
$5x + 2y \leqslant 80$	0.5 (machine-hours)
$3x + 4y \leqslant 90$	0.5 (man-hours)

in the objective function) he should accept because by using it his profit can increase by 0.5. If he has to pay an extra amount for the extra unit, 0.5 is the maximum he should pay. An extra man-hour (91 available instead of 90) would also increase the objective function value by 0.5 to 85.5. It is only by chance that the dual values are equal. Usually the dual values will differ from constraint to constraint. Dual values are measures of the true worth to the decision-maker of the resources. They are likely to differ from the price paid for the resource and its market value. The dual values are produced as part of the output of an algebraic solution of an LP problem and require no extra calculations once the LP has been solved.

Dual values are sometimes equal to zero. In the example, both constraints were *tight*, meaning that the total availability of each resource was used up. (Check this by substituting the optimal x, y values in the constraints). In some circumstances the total availability of a resource might not be fully used up, as would here have been the case if there had been more constraints. If not all the resource is used up the constraint is said to be *slack* and the associated dual value (of that constraint only) must be zero, since available units of that resource are already not being employed and therefore extra units can have no value to the decision-maker. The slack for any constraint is the quantity of that particular resource that is left unused. If the optimal solution revealed, for example, that only 50 man-hours were used then the slack for that constraint would be 40 (= 90 − 50). When the slack is non-zero then the dual value is zero.

A dual value can be calculated graphically by reworking the solution with the right-hand side of the constraint increased by 1 (see figure 9.10). The amount by which the objective function has

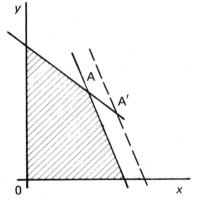

Figure 9.10

increased in the new solution is the dual value. In the example, the dual value of machinery can be calculated by solving the problem as before, but with the machinery constraint being $5x + 2y \leqslant 81$ instead of $5x + 2y \leqslant 80$. In this case, the optimal point is the point of intersection of:

$$5x + 2y = 81$$

$$3x + 4y = 90$$

The optimal point is no longer A, but is A^1. The x and y values can be calculated by simultaneous equations as before, giving:

$$x = 10\tfrac{2}{7} \qquad y = 14\tfrac{11}{14}$$

The objective function value at this point is:

$$4(10\tfrac{2}{7}) + 3(14\tfrac{11}{14}) = 85.5$$

The dual value of machine-hours is the increase in the objective function:

$$\text{dual value} = 85.5 - 85$$

$$= \underline{0.5}$$

At first sight it may seem paradoxical that an extra machine-hour could be used at all. In the original solution all 80 machine-hours and all 90 man-hours were being used. Since both products X and Y require a combination of the resources, how could the extra machine-hour be used since there are no man-hours with which to combine it? The answer is that the production mix changes. Fewer units of Y are made (y decreases from 15 to $14\tfrac{11}{14}$). This frees up some units of both resources which can be combined with the extra machine-hour to produce more of product X (x increases from 10 to $10\tfrac{2}{7}$). The new production mix is more profitable by 0.5 as measured by the dual value.

Dual values are marginal values and are not guaranteed to hold over large ranges. If the decision-maker were offered 10 extra machine-hours at an additional price of 0.3, should he accept all 10? Or is the range over which the dual value holds less than 10? Most computer packages give the range for which the dual value holds. This is known as the *right-hand side range*.

Note also that the dual value works in both directions. A dual value of 0.5 also means that one unit *less* of the resource would *lower* the objective function by 0.5.

Interpreting computer output

In practice, LP problems are solved by computer. Many LP packages are interactive and the details of the problem are put into the computer in response to a series of questions. The output includes details of the solution and dual values.

EXAMPLE

A manufacturer of soft drinks is preparing a production run on two drinks, A and B. There are sufficient ingredients available to make 4000 bottles of A and 8000 of B, but there are only 9000 bottles into which either may be put. It takes 30 minutes to prepare the ingredients to fill 200 bottles of A; it takes 10 minutes to prepare the ingredients to fill 200 bottles of B. There are 12 hours available for this operation. The profit is 8p/bottle for A and 6p/bottle for B.

Formulate this problem as an LP. Solve by computer and interpret the output.

Isolate the three elements of an LP problem:

(a) *Decision variables*

The manufacturer wishes to know how many bottles of each soft drink he should prepare. Therefore the decision variables must be:

$$x = \text{bottles of drink A}$$
$$y = \text{bottles of drink B}.$$

(b) *Objective function*

The objective is to maximize profit. The objective function (in £) is:

$$0.08x + 0.06y$$

(c) *Constraints*

The ingredients to make each type of drink are limited. The first two constraints express these limitations:

$$x \leqslant 4000 \text{ (for drink A)}$$
$$y \leqslant 8000 \text{ (for drink B)}.$$

The total number of usable bottles is also limited. Since the bottles are suitable for either A or B,

$$x + y \leqslant 9000 \text{ (bottle availability)}$$

The next constraint concerns the time availability on the machine which fills the bottles. For each bottle of A, the filling time is 30/200 minutes. For each bottle of B, the filling time is 10/200 minutes. Therefore, the total time required on the machine for both types of drink is:

$$\frac{30}{200} x + \frac{10}{200} y \leqslant 12 \times 60$$

Tidying up this equation, it becomes:

$$30x + 10y \leqslant 12 \times 60 \times 200 \quad \text{(multiplying by 200)}$$

$$3x + y \leqslant 14,400 \quad\quad\quad \text{(dividing by 10)}$$

Lastly, restrict the values of the decision variables to be non-negative:

$$x, y \geqslant 0$$

The fully formulated LP is then:

$$\text{Max.} \quad 0.08x + 0.06y$$
$$\text{Subject to} \quad x \leqslant 4000$$
$$y \leqslant 8000$$
$$x + y \leqslant 9000$$
$$3x + y \leqslant 14,400$$
$$x, y \geqslant \quad\quad 0$$

A typical computer package used to solve this problem would give output as in table 9.4. The output can be interpreted as follows:

(1) The optimal production levels are to make 2700 bottles of drink A and 6300 of B.

(2) At these production levels the profit will be £594 (check this by substituting $x = 2700$ and $y = 6300$ into the objective function).

(3) The column headed 'slack' tells whether each constraint is tight or slack. If a constraint is slack the extent of the spare capacity is given by the figure in the column. For example, the production of soft drink A is not restricted by the availability of ingredients (constraint

Table 9.4   Optimal solution

Variable	Value
x	2700
y	6300
Objective value = 594	

Constraint	Slack	Dual
1	1300	0
2	1700	0
3	0	0.05
4	0	0.01

1) of which there are sufficient to make a further 1300 bottles. The slack values may be checked by substituting $x = 2700$ and $y = 6300$ into each of the constraints, e.g. for the first constraint, put $x = 2700$: $2700 + \text{Slack} = 4000$.

(4) The column headed 'Dual' gives the dual values or shadow prices for each resource or constraint. As expected a slack constraint has a zero dual value. This is the case for constraints 1 and 2 referring to the availability of bottles. The third constraint is tight and has a dual value of 0.05. If an extra bottle were available, 5p could be added to total profit. The fourth constraint is also tight, with a dual value of 0.01. The meaning of this dual value is harder to interpret. What is it that is worth an extra 1p? An extra hour, an extra minute? It is necessary to go back to the constraint itself: $3x + y \leqslant 14{,}400$. The dual value refers to the increase in the objective function if the right-hand side of the constraint were increased by 1 from 14,400 to 14,401. Originally this constraint was that the time available on the machine should be less than 12 hours. When the constraint was formulated, hours were turned into minutes so that the right-hand side was 12 x 60. To get rid of some fractions the constraint was multiplied through by 20 so that the right-hand side became 12 x 60 x 20 = 14,400. The units are therefore one-twentieth of a minute, i.e. 3 seconds. The dual value means that if extra time were available on the machine each extra 3 seconds would be worth 1p. If the extra time could be gained by use of overtime then each 3 seconds would

be worth 1p, each minute would be worth 20p, each hour would be worth 1200p, or £12. It is worth using overtime if the cost is less than £12/hour, over and above the cost of normal-time working included in the objective function.

Extensions of linear programming

Linear programming as described here is capable of solving a wide range of practical problems. The basic technique, however, can be extended in a number of directions to cover an even greater range.

The first extension is the handling of *minimization* as well as maximization problems. Usually minimization problems have objective functions in terms of costs. The formulation of such problems is the same, except that the minimization constraints tend to be 'greater than or equal to' rather than 'less than or equal to', but are not exclusively so. Graphical solution proceeds by drawing the feasible region and then moving the objective function lines ('isocost' lines) in a south-westerly (as opposed to north-easterly) direction to find the optimal point. The second worked example below is a minimization problem.

A second extension is to *integer programming*, where some or all the variables are restricted to take integer (whole-number) values only. This would be valuable in a case like the following. A company has a series of opportunities for capital investment which have different returns and cash flow profiles. Only a limited amount of money is, of course, available and also the company has certain cash-flow requirements through time. In which of the opportunities should the company invest? The problem can be handled by LP. The decision variables are the amounts of money to invest in each opportunity; the objective is to maximize return; the constraints are the money available and the cash-flow requirements through time. Some of the opportunities can have variable amounts invested in them, such as for example the purchase of stocks and shares. Other opportunities, however, require all-or-nothing investments, such as the purchase of a motor vehicle or the construction of a new plant. An answer, which was to invest £20,000 in new vehicles when the vehicles cost £15,000 each, would not make sense. The investment must be for £0, £15,000, £30,000, etc. This situation requires a variable defined as the number of vehicles to buy which can take on whole number values only, i.e. 0, 1, 2, 3 . . . . Integer programming allows some variables to be of this type. As with basic LP, computer packages are available to solve this type of problem. Since the method of solution is more complicated, it is often extremely time-consuming to solve large problems via integer programming.

A third extension to basic LP is to *general mathematical programming*. This term refers to formulations where the linearity requirement is relaxed. For example, the objective function may involve squared terms, allowing for a curved relationship between profitability and output. Only a few types of non-linear problem can be solved, and those that can be solved generally require a large computer capacity.

Integer and mathematical programming have been mentioned to illustrate the ways in which LP can be extended to other situations. These areas will not be pursued here. In practice only small problems of this type (in terms of numbers of decision variables and constraints) are solved.

Worked examples

1. P. E. WOOD

The company produces chairs and tables which must be processed through assembly and finishing departments. Production of a chair requires 4 hours of assembly time and 2 hours of finishing time. The corresponding times for a table are 2 hours and 5 hours. There are 80 hours of assembly time available per week and 120 hours of finishing time. The profit contribution (revenue − variable cost) for a chair is £30 and for a table £40. What production levels per week will maximize total profit contribution? Extra finishing time can be made available by using overtime, which will cost £4 per hour in addition to normal payment. Should overtime be used?

(a) *Decision variables*
Let x = number of chairs manufactured.

Let y = number of tables manufactured.

(b) *Objective function*
Maximize contribution:

   max $30x + 40y$.

(c) *Constraints*

$$4x + 2y \leqslant 80 \text{ assembly time}$$

$$2x + 5y \leqslant 120 \text{ finishing time}$$

$$x, y \geqslant 0.$$

The graphical solution is shown in figure 9.11. The optimal point is A at the intersection of the assembly and finishing constraints. The x and y values can be read off from the graph or found from the simultaneous equations.

$$4x + 2y = 80 \tag{1}$$

$$2x + 5y = 120 \tag{2}$$

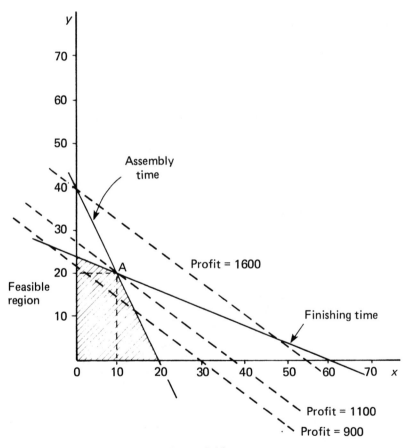

Figure 9.11

Multiply (2) by 2 and subtract (1)

$$4x + 10y = 240$$
$$4x + 2y = 80$$

$$8y = 160$$
$$y = 20$$

Substituting $y = 20$ in (1),

$$4x + 40 = 80$$
$$x = 10$$

Point A is optimal solution: $x = 10$ chairs

$$y = 20 \text{ tables}$$

Substitute in the objective function

Objective value = 30 x 10 + 40 x 20

$$= £1100$$

If an extra hour of finishing time were available, the optimal point would be A^1 instead of A (see figure 9.12) and its coordinates would

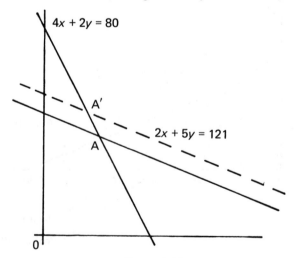

Figure 9.12

be given by the two equations:

$$4x + 2y = 80$$
$$2x + 5y = 121$$

Multiply the second equation by two and subtract the first:

$$4x + 10y = 242$$
$$\underline{4x + 2y = 80}$$
$$8y = 162$$
$$\underline{y = 20.25}$$

Substitute $y = 20.25$ in either equation,

$$4x + 2(20.25) = 80$$
$$4x = 39.5$$
$$\underline{x = 9.875}$$

The new optimum is 20.25 tables, 9.875 chairs, so the extra time allows output of the more profitable product (tables) to be produced. It is assumed 0.25 table and 0.875 chair refer to partly finished products.

The new objective function value is:

$$30 \times 9.875 + 40 \times 20.25$$
$$= 296.25 + 810$$
$$= 1106.25$$

The dual value of finishing time is therefore

$$\text{New objective value} - \text{Old objective value} = 1106.25 - 1100$$
$$\underline{= 6.25}$$

An extra hour of finishing time increases the contribution by £6.25. The normal cost of finishing time is already included in the objective function which measures contribution not revenue. The dual value (£6.25) should be compared with the extra overtime cost of finishing time (£4). Since it is greater, then overtime should be used.

2. B. A. ALCORN

Farmer Alcorn has to decide the correct amount of fertilizer to apply to his root-vegetable fields. He has been advised by a soil analyst that each acre will require the following minimum amounts: 60 lb nitrogen compounds, 24 lb phosphorus compounds and 40 lb potassium compounds. There are two major types of fertilizer available. The first is JDJ which comes in 40 lb bags at £6 per bag and which comprises 20% nitrogen compounds, 5% phosphorus and 20% potassium. The second, PRP, comes in 60 lb bags at £5 each and comprises a 10%–10%–5% mixture. Formulate the problem as an LP which minimizes cost.

How many bags of each product should he use per acre?

Being a cautious farmer he wishes to have a second opinion from another soil analyst. If the second analyst suggests 9 lb less phosphorus compounds are required per acre, how much will the farmer save per acre? What is the dual value for phosphorus?

Note that this is a minimization problem:

(a) *Decision variables*

Let x = number of bags of JDJ.

Let y = number of bags of PRP.

(b) *Objective function*

The objective is to minimize total costs:

$$\min 6x + 5y.$$

(c) *Constraints*

There are constraints for the requirements of each of nitrogen, phosphorus and potassium.

If x bags of JDJ are used, then $40x$ lb of JDJ are used. Since JDJ contains 20% nitrogen compounds, the quality of nitrogen applied will be 20% of $40x$, i.e. $8x$ lb.

Similarly, the nitrogen in y bags of PRP is 10% of $60y$, i.e. $6y$ lb. The total amount of nitrogen applied will therefore be:

$$8x + 6y$$

This must be greater than the amount required which is 60 lb. The constraint must be:

$$8x + 6y \geqslant 60.$$

The other two constraints can be worked out in the same way:

$$2x + 6y \geqslant 24 \text{ (phosphorus)}$$

$$8x + 3y \geqslant 40 \text{ (potassium)}$$

Non-negativity constraints are also required.

The complete formulation is:

$$\text{Min } 6x + 5y$$

$$\text{Subject to } 8x + 6y \geqslant 60 \quad \text{Constraint 1}$$

$$2x + 6y \geqslant 24 \quad \text{Constraint 2}$$

$$8x + 3y \geqslant 40 \quad \text{Constraint 3}$$

$$x, y \geqslant 0$$

The graphical solution is shown in figure 9.13. Note that when a constraint is 'greater than', the feasible area defined by it is on the opposite side of the line than for a 'less than' constraint. Note also that the objective line is moved downwards in the minimizing process. A carefully drawn graph shows that the optimal point is A, at the intersection of constraints 1 and 2. Its coordinates are given by the simultaneous equations:

$$8x + 6y = 60$$

$$2x + 6y = 24$$

Subtract them:

$$6x = 36 \quad \underline{x = 6 \text{ bags}}$$

Substitute $x = 6$

$$8 \times 6 + 6y = 60 \quad \underline{y = 2 \text{ bags}}$$

At this point the objective function value is:

$$(6 \times 6) + (5 \times 2) = £46$$

The answer to the problem is to use 6 bags of JDJ and 2 bags of PRP per acre at a cost of £46/acre.

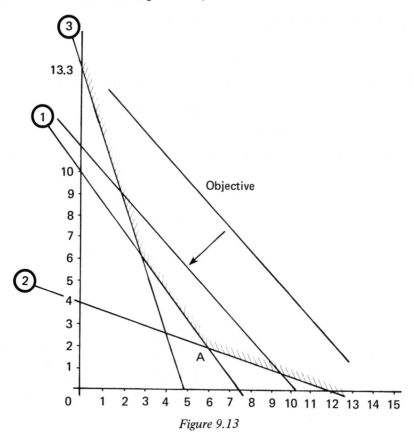

Figure 9.13

If 9 lb less of phosphorus are required the associated constraint (2) will be $2x + 6y \geqslant 15$ and the graphical solution will be as in figure 9.14. The optimal point is B, still at the intersection of constraints 1 and 2, but also on the x axis. It can be seen from figure 9.14 that at B,

$$x = 7\tfrac{1}{2}, \quad y = 0.$$

The new objective function value is $(6 \times 7\tfrac{1}{2}) + (5 \times 0) = £45$. The farmer will therefore save £1/acre if the second analyst's advice is thought to be correct.

The dual value for phosphorus is the change in the objective function if the requirement for phosphorus is changed by 1. It has not been changed by 1 but by 9 (from 24 to 15) in which case the objective function changed by £1. The dual value of phosphorus

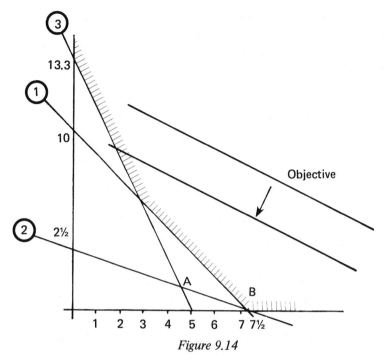

Figure 9.14

must therefore be:

$$\frac{£1}{9} = \underline{0.11p}$$

3. A. C. SAWYER

The company wants to make the best use of its wood resources in one of its forest regions which contains fir and spruce trees. During the next month 45,000 board-feet of spruce are available and 80,000 board-feet of fir.

The company has mills to produce lumber or plywood. Producing 1000 board-feet of lumber requires 1200 board-feet of spruce and 2500 board-feet of fir. Producing 1000 square feet of plywood requires 2200 board-feet of spruce and 3500 board-feet of fir.

Firm orders mean that at least 6000 board-feet of lumber and 12,000 board-feet of plywood must be produced during the next month. The profit contributions are £125 per 1000 board-feet of lumber and £225 per 1000 square feet of plywood.

The production decision is taken on the basis of the following LP model:

$$\text{Max } 125L + 225P$$

$$\text{Subject to } 1.2L + 2.2P \leqslant 45 \text{ (spruce)}$$

$$2.5L + 3.5P \leqslant 80 \text{ (fir)}$$

$$L \geqslant 6 \text{ (orders)}$$

$$P \geqslant 12 \text{ (orders)}$$

where L = amount of lumber produced (in '000 board-feet)

P = amount of plywood produced (in '000 square feet).

The computer output for this problem is of the form shown in table 9.5. Assuming the constraints are in the same order as in the formulation (e.g. constraint 1 refers to spruce):

(a) How much lumber and how much plywood should be produced?
(b) What profit contribution should result from this production mix?
(c) What numbers should be where the question marks are?
(d) How much extra profit contribution could result if the amount of spruce that arrives at the mills is found to be 45,500 board-feet?

Table 9.5

Variable	Value
L	14.23
P	12.69

Objective value = 4634

Constraint	Slack	Dual price
1	0	96.15
2	0	3.85
3	?	0
4	?	0

(a) The optimal values of the variables are given as 14.23 and 12.69. Therefore <u>14,230 board-feet</u> of lumber should be produced and 12,690 square feet of plywood.

(b) The optimal objective value is £4634, which is the profit contribution at the production levels given in (a) above. This can be checked by substituting $L = 14.23$ and $P = 12.69$ in $125\,L + 225P$.

(c) The question marks are in the positions of the slacks for the two minimum order level constraints.

For $L \geqslant 6$, the slack is $14.23 - 6 = \underline{8.23}$

For $P \geqslant 12$, the slack is $12.69 - 12 = \underline{0.69}$

Note that for $\geqslant$ constraints it is more usual to refer to these amounts as 'surplus' not 'slack'.

(d) The dual value for constraint 1 (referring to spruce) is £96.15. This means that an extra 1000 board-feet of spruce would produce a further £96.15 in profit contribution. An extra 500 board-feet could therefore be expected to produce an extra <u>£48.075</u>.

Final comments

From the management point of view it is important to regard LP as an *aid* to decision-making. It would be a mistake to think of LP as a substitute for decision-making. LP is just one of the factors that may contribute to a good decision being taken. It would be equally a mistake to suppose that, because LP may not be able to capture all aspects of a problem, then it cannot be used at all and the problem should be tackled by instinct. A balanced approach would be to use LP to deal with quantifiable parts of the problem and build a sound basis on which to build an evaluation of the qualitative parts.

LP contributes to a decision problem by solving it and by providing a range of economic information about it. It also makes a less direct but equally important contribution because the formulation of the problem as an LP in itself gives a greater understanding of the problem's structure and the range of alternative actions available.

Examples of LP are usually to do with products and resources, but the technique can be applied to any situation that has the key characteristics which indicate an LP. To summarize, an LP is recognized by looking for the three factors:

- decision variables
- linear objective function
- linear constraints.

Further examples

1. T. E. M. PENMAN

This office equipment company produces two types of desks: Standard and Executive. These desks are sold to a large office furniture retailer, and for all practical purposes there is an unlimited market for any mix of these desks, at least within the manufacturer's production capacity. Each desk has to go through four basic operations: cutting the lumber, joining the pieces, prefinishing, and final finish. Each unit of the Standard desk produced takes:

48 minutes of cutting time;
2 hours of joining;
40 minutes of prefinishing;
5 hours 20 minutes of final finishing time.

Each unit of the Executive desk requires:

72 minutes of cutting;
3 hours of joining;
2 hours of prefinishing;
4 hours of final finishing time.

The daily capacity for each operation amounts to:

16 hours of cutting;
30 hours of joining;
16 hours of prefinishing;
64 hours of final finishing time.

The profit per unit produced is:

£40 for the Standard desk;
£50 for the Executive desk.

(a) Formulate this problem as a linear programme with the objective of maximizing daily profit.

(b) Find the optimal solution graphically. What is the amount of slack for each constraint?
(c) Find the shadow price for each constraint and interpret its meaning.

2. G. R. RUFF

A pet food is to be made out of sago flour, meat meal and enriched skimmed milk. The cheapest formula that will supply a prescribed amount of protein and fat and satisfy a restriction on the upper limit on the combined total of fibre and ash is to be chosen. The chemical analyses and costs of the three possible ingredients, together with dietary requirements, are given in table 9.6. The figures given (including the requirements) are percentages of dry weight.

Formulate the problem as a linear programme.

Table 9.6

	Requirement	Sago flour	Meat meal	Enriched skimmed milk
Protein	≥ 20	0	50	30
Fat	≥ 10	0	10	20
Carbohydrate	Any	100	0	40
Fibre and ash	≤ 20	0	40	10
Cost (£ per ton)		40	50	70

3. S. I. POTTS

The company produces and sells two brands of tea: Choice and Standard. To preserve their market position they control quality as closely as possible. Quality is primarily a combination of two characteristics, 'body' and 'flavour'. For the Choice brand the 'flavour' index must be equal to 15 and the 'body' index no greater than 9. For the Standard brand the flavour index must be at least 13 and the body index should not be greater than 10. Potts are developing an order for tea leaves to meet expected sales requirements of at least 20,000 lb of the Choice brand and at least 50,000 lb of the Standard brand at the end of the processing period. The company

Table 9.7

	Patel	Singh	Anand
Body index	9	11	8
Flavour index	17	12	14
Pounds available	30,000	45,000	60,000
Cost per pound	£0.80	£0.56	£0.64

buys from three brokers: Mr Patel, Mr Singh and Mr Anand who have provided information about their available stocks (table 9.7). Formulate this linear programming problem.

ANSWERS

1. *T. E. M. Penman (see figure 9.15)*

Max $40S + 50E$

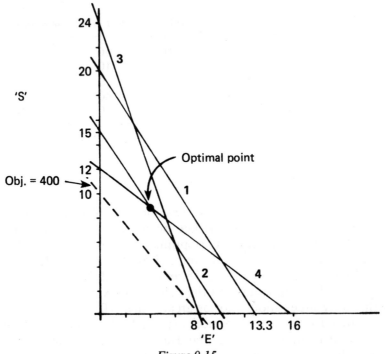

Figure 9.15

s.t. 1 $\frac{4}{5}S + \frac{6}{5}E \leqslant 16$ cutting time (hours)

2 $2S + 3E \leqslant 30$ joining time (hours)

3 $\frac{2}{3}S + 2E \leqslant 16$ prefinish time (hours)

4 $\frac{16}{3}S + 4E \leqslant 64$ final time (hours)

Optimal solution: S = 9, E = 4   Obj. = 560

Constraint	Slack	Dual
1	4	0
2	0	40/3
3	2	0
4	0	2½

2. *G. R. Ruff*

Let x_1 = proportion of sago flour in diet;

x_2 = proportion of meat meal in diet;

x_3 = proportion of skimmed milk in diet.

$$\text{Min } 40x_1 + 50x_2 + 70x_3$$
$$\text{Subject to } 50x_2 + 30x_3 \geqslant 20$$
$$10x_2 + 20x_3 \geqslant 10$$
$$40x_2 + 10x_3 \leqslant 20$$
$$x_1 + x_2 + x_3 = 1$$
$$x_1, \ x_2, \ x_3 \geqslant 0$$

3. *S. I. Potts*

The variables can be represented as x_{ij}

where i = 1, 2, 3 (Patel, Singh, Anand)

j = 1, 2   (Choice, Standard respectively)

e.g. x_{32} = lb of Anand tea put into the Standard brand.

Objective function
The objective is to minimize cost:

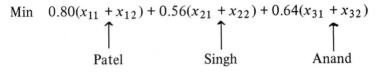

$$\text{Min} \quad 0.80(x_{11} + x_{12}) + 0.56(x_{21} + x_{22}) + 0.64(x_{31} + x_{32})$$

| Patel | Singh | Anand |

Constraints
The constraints reflect the requirements for the brands in terms of flavour and body, the minimum expected sales and the availability of supply.

(1) $17x_{11} + 12x_{21} + 14x_{31} = 15(x_{11} + x_{21} + x_{31})$ Choice flavour

(2) $17x_{12} + 12x_{22} + 14x_{32} \geqslant 13(x_{12} + x_{22} + x_{32})$ Standard flavour

(3) $9x_{11} + 11x_{21} + 8x_{31} \leqslant 9(x_{11} + x_{21} + x_{31})$ Choice body

(4) $9x_{12} + 11x_{22} + 8x_{32} \leqslant 10(x_{12} + x_{22} + x_{32})$ Standard body

(5) $x_{11} + x_{21} + x_{31} \geqslant 20{,}000$ Choice sales

(6) $x_{12} + x_{22} + x_{32} \geqslant 50{,}000$ Standard sales

(7) $x_{11} + x_{12} \leqslant 30{,}000$ Availability Patel

(8) $x_{21} + x_{22} \leqslant 45{,}000$ Availability Singh

(9) $x_{31} + x_{32} \leqslant 60{,}000$ Availability Anand

The first four constraints need to be rearranged:

(1) $\quad 2x_{11} - 3x_{21} - x_{31} = 0$

(2) $\quad 4x_{12} - x_{22} + x_{32} \geqslant 0$

(3) $\quad \phantom{4x_{12}} - 2x_{21} + x_{31} \geqslant 0$

(4) $\quad x_{12} - x_{22} + 2x_{32} \geqslant 0$

Simulation

By the end of the chapter the reader should know what a simulation is, what are the key factors in its use and what are its drawbacks. The chapter is about using, not constructing, simulations. The latter almost certainly would require computer programming expertise.

Simulation means the imitation of the operation of a system. Outside the business world, simulations may be encountered as wind tunnels for estimating the performance of newly designed aircraft and flight simulators for testing the reactions of aircraft pilots; in business a simulation may replicate cash flows in a company to demonstrate the impact on profitability of changing circumstances or it may imitate the inventory system to show how different policies may affect costs and the likelihood of being out of stock. In the case of a wind tunnel, the simulation is based on a small-scale physical model. Physical models may also be used in business applications (for example in planning the layout of a new supermarket), but more usually the simulation is based on a series of mathematical equations. Large simulations of this second type are usually computerized.

A simple example of a mathematical simulation might be the

calculation of the cash flows from alternative capital expenditure projects. Suppose projects have the following characteristics in common:

- turnover generated in 1st year = £2 million;
- operating costs = 60% of turnover;
- overheads p.a. = £500,000;
- length of project = 3 years;

The three alternative projects A, B and C differ in the initial capital outlay and the rate at which turnover increases over the 13 years of the project. Given these figures, a cash flow for each project can be calculated. The calculation process is called a *model*. Using the model is a *simulation* (figure 10.1). Given the capital cost and the turnover increase percentage, the model imitates the annual cash flows to produce a total cash flow for a project.

The purpose of a simulation is to test the effect of different *decisions* and different *assumptions*. In the above example the different decisions are the three projects; the different assumptions are the first year turnover, the relationship between costs and turnover, the level of overheads, etc.

The advantage of using simulation is that different decisions and assumptions can be tested out quickly and without the expense or danger (in the case of a flight simulator) of carrying out the decision in reality. Its disadvantage is that it is completely uncreative. Like a large calculator, it merely measures the effect of a decision defined by the operator. The 'best' decision means the best of those that have been tested. There may well be other better decisions that the operators of the simulation have not thought of. In the capital project example there may be another project better than any of the three

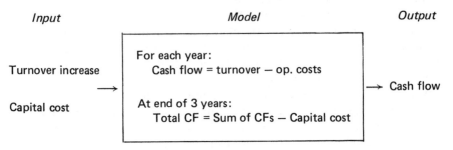

Figure 10.1

tested. The simulation will not indicate that there is a best project. It will merely measure the cash flows of the projects fed into it. Compare this with the optimization techniques such as linear programming and decision analysis which actually themselves define the best decision. Provided the decision-maker accurately models the problem, LP or decision analysis will come up with the optimal decision. In simulation, the decision-maker must not only model the problem accurately but also must propose the whole range of possible decisions.

Simulation is, therefore, a different method of contributing to decision-making than are the optimization techniques. Its great advantage is that it is capable of a far wider range of applications. The optimization techniques can only be applied in particular well-defined circumstances. For instance, LP applies only when the problem can be formulated with an objective function and constraints expressed linearly in terms of continuous variables; decision analysis is applicable only when the problem can be formulated as a series of subdecisions and chance events represented by a decision tree. No such restrictions limit the uses of simulation. Indeed, the widespread use of simulation models stems from the fact that there are so many situations which cannot be contained within the underlying assumptions of optimization techniques (or rather the situations must be approximated and simplified to an unrealistic extent), but which can be explored using simulation.

Application of simulation to a distribution system

Because it is not restricted to problems with particular characteristics, simulation has a wide range of applications, including corporate planning models, business games, and models of operations. The application described here is the distribution of steel throughout a country.

PROBLEM

On a South East Asian island (see figure 10.2) a new large steelworks was being built at location X. At the time all steel was imported, but upon completion of the plant all steel required on the island was to be supplied by it. Demand could be thought of as arising at 120

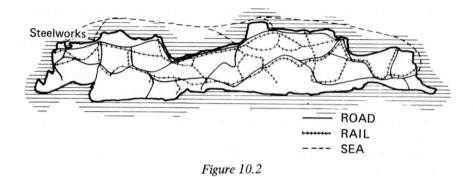

Steelworks

——— ROAD
⊢⊣⊢⊣ RAIL
- - - - SEA

Figure 10.2

centres on the island. The problem was to decide the best distribution network for the steel products. Three modes of transport were available. Sea routes for longer distances were viable because of the lack of infrastructure, although suitable facilities were available at only one or two harbours; rail routes existed but they were radial as a result of their original purpose which was (and is) the transport of tea from the growing areas in the mountainous centre to the population and harbours of the coast; road routes covered most of the island but few of them had a tarmac surface.

SIMULATION

The size of the problem (different products, different modes of transport, 120 demand centres), the fact that some possible solutions required large capital expenditures (in harbour facilities, transit warehouses) and the many uncertainties (not least the levels of demand) meant that no optimization technique was able to provide a solution. A simulation was used. It specified all the links in the network and the costs associated with transporting steel products across the links by each of the transport modes. It calculated the total tonnage moving along each link, checked that this did not exceed capacity and then calculated the total cost of supplying demand under a given distribution policy. The policies tested were the sets of routes by which each demand centre was to be supplied, together with the associated capital investments. There were a very large number of possible policies and only a few could be tested; therefore, as with all simulations, the finally selected policy was at best an approximation to the optimum. Because there were so many possible policies it was essential to adopt a structured approach to selecting a 'best' policy. At each stage of the investigation the range of possibilities had to be narrowed down. The policies tried out first were:

1 Supply all demand by road.
2 Supply all demand possible by rail, the remainder by road.
3 Supply all demand possible by sea, the remainder by road.
4 Supply all centres more than 1000 miles away by sea; between 500 and 1000 miles by rail; less than 500 miles by road.

The results from these four policies, together with the costs associated with individual links, indicated the types of policy worthy of further investigation. The process of narrowing down continued. At each stage some policies were rejected, others amended and refined. Finally, a satisfactory policy was evolved.

The purpose of a simulation is to evaluate decisions and assumptions. In this case the decisions are the different distribution policies, the assumptions are the levels of demand in the 120 centres, the capital costs, the transportation costs and capacities of the individual links. By varying the assumptions, the ability of the chosen policy to stand up to changing circumstances was investigated.

Types of simulation

The application described here is a large one requiring a considerable computer program to make the calculations. It is an example of the simulation of a *physical system*. Although the end-result is financial, the simulation is based on volume flows down the various links. Contrast this with a simulation of a *financial system* based purely on cashflows.

As well as having a physical or financial application, simulations are also classified as *deterministic* or *stochastic* (see figure 10.3). Deterministic means that the inputs and assumptions are fixed and known. None are subject to probability distributions. The simulation acts as a large calculating mechanism. Simulation is used because of the number and complexity of the relationships involved. Corporate planning models are often examples of this type. The impact on profit of different assumptions concerning, say, the growth of sales, are calculated by the simulation model. Associated decisions relating

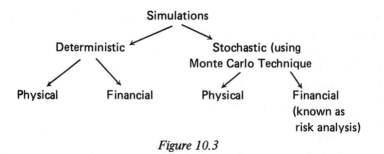

Figure 10.3

to, perhaps, production capacity or advertising can also be tried out and eventually plans for the future formulated.

Stochastic means that some of the inputs, assumptions or variables are subject to probability distributions. In such a case the simulation is run many times (hence the need for a computer) with the stochastic variables taking on different values for each run. But it is done in such a way that the different values used have a distribution which is the same as the actual distribution for that particular variable. This technique is known as *Monte Carlo simulation*. Stochastic financial simulations are often known as *risk analyses*. An application will be described later.

MONTE CARLO TECHNIQUE

The Monte Carlo technique will be illustrated in the context of a simple problem.

Raw materials and sub-assemblies used in a manufacturing process are stored in a warehouse. Each day the foreman of the process requisitions items from the warehouse. The demand for each item varies from day to day. If the warehouse can supply the items, it does. If there is an item which is out of stock the foreman has to improvise in order to keep the manufacturing process going. In spite of long experience as an improviser, this is usually done at some cost to the company. It also costs the company money to hold stock. For this reason, the warehouse does not keep vast stocks, but from time to time when the stock is low re-orders each item from suppliers. The time taken for the replenishment to arrive varies from item to item and from order to order.

The company is now trying to define its best inventory policy. It is starting by looking at one item, data about which have been collected from stores records and from the accounting department.

Table 10.1 Demand distribution

Table 10.1 Demand distribution

Demand	Probability
1	0.10
2	0.15
3	0.25
4	0.25
5	0.15
6	0.10
	Total 1.00

Table 10.1 shows the demand for the item varies from 1 to 6 per day. Table 10.2 shows that when the item is re-ordered it takes from 4 to 7 days to arrive at the warehouse. Table 10.3 shows that the cost of holding an item of stock for 1 day is £1. The cost of not being able to supply an item requisitioned is £20.

The company is trying to find the best inventory policy for this item. By inventory policy is meant the re-order quantity and the re-order level (the level of stock which triggers a replenishment

Table 10.2 Order lead-time

Lead-time (days)	Probability
4	0.1
5	0.4
6	0.4
7	0.1
	Total 1.0

Table 10.3 Costs

Stockholding cost per item per day = £1

Lost demand/item/day = £20

order). A simulation is to be used which will model the operation of the warehouse over 40 days and calculate the cost of running a particular inventory policy over that time.

Obviously, the daily demand is an input to the simulation. The average value of demand (3.5 units/day) cannot be used because this may mask the peak and trough effects. A different level of demand must be used each day, but at the end of the period the distribution of daily demand must be almost the same as in table 10.1. A second characteristic of the process is that the different daily demands must occur in random order. For example a simulation over 100 days which had a demand for 1/day for the first 10 days, a demand 2/day for the next 15 days, etc., would ensure that the distributions matched but would not be in accord with what would happen in practice, and therefore would not adequately test alternative inventory policies.

The Monte Carlo technique is a method of providing inputs to the simulation which satisfy both characteristics. First some random numbers are required (see table 10.4). Random numbers are just what they appear: a series of numbers in random order. They may come from a computer-generating procedure or from a table found

Table 10.4   Random numbers

97	51	88	70	96	77	96	76	18	54
80	94	76	58	70	12	58	63	05	13
14	02	38	66	86	96	91	42	60	63
22	52	78	18	24	77	87	24	08	06
96	44	82	84	70	10	08	68	90	76
49	15	57	51	92	14	67	83	42	07

Table 10.5   Demand distribution

Demand	Probability	Associated random numbers
1	0.10	00–09
2	0.15	10–24
3	0.25	25–49
4	0.25	50–74
5	0.15	75–89
6	0.10	90–99
	1.00	

at the back of an operational research textbook. Second, numbers are associated with the different levels of demand, as in table 10.5. The random numbers in table 10.2 all have two digits. The 10 numbers (00–09) are associated with a demand of 1 unit/day, 15 numbers (10–24) with a demand of 2 units/day, 25 numbers (25–49) with a demand of 3 units/day, etc. Whenever a demand level is required in the simulation a random number is taken and the associated demand level used in the simulation. Going through the random number table each demand level has the same chance of occurring as in the original distribution; e.g. there is a 10% chance of a random number in the range 00–09, therefore there is a 10% chance of using a demand of 1 unit/day in the simulation. Moreover, the demands used will be in random order.

The first random number is 97, so the demand for Day 1 of the simulation will be 6 since 97 is in the range 90–99 which is associated with a demand for 6. The second random number is 51 so the second day's demand will be 4. In this way a stream of daily demands are provided for the simulation which, over a long run, will match what happens in practice.

FLOWCHARTS

The inventory control simulation involves a fairly simple process. In practice situations, such as the steel distribution problem, may be complex. A flowchart is an intermediate stage which helps transform a verbal description of a problem into a simulation. It is merely a diagrammatic representation of the steps that have to be followed in the simulation. It includes the decisions and actions that are carried out in operating the actual inventory system as well as the extra steps of recording information during the simulation. In a flowchart, these symbols are used:

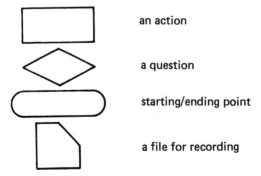

an action

a question

starting/ending point

a file for recording

Solid lines show the flow of events through time; dotted lines represent movements of information. Figure 10.4 shows the flowchart for the inventory problem.

The first step each day is to see whether new supplies will arrive that day. This requires information from the replenishment file which stores details of orders in the system that have yet to arrive. If new supplies are to arrive then the replenishment file and the inventory file (showing the current level of stock) must be updated; if not the simulation moves on to the next step, which is to generate the days demand by the Monte Carlo method.

The question is then asked: is the demand greater than current inventory? This requires information from the inventory file. If demand can be supplied, parts are moved from the store and the inventory file is updated; if not, as much of the demand as possible is supplied and the rest is recorded as lost demand in the lost demand file.

The next step is to calculate costs. In this example there are just two kinds of cost, the cost of holding inventory and the cost of losing demand. Both are calculated from the appropriate file information.

Then the end-of-day inventory level is inspected to see if it has reached the point at which a new order should be placed. If so, the replenishment file records the placing of the new order. Lastly the simulation advances to a new day.

A flowchart, therefore, describes the logic of the simulation and the way in which it attempts to imitate the real system. There are other ways of representing simulations diagrammatically, but flowcharts are the most common. The symbols used in the flowchart are not standard. Other flowcharts may use different symbols. Whatever the style of representation, all these diagrams have the same purpose, which is to describe the logical steps of the simulation as a preliminary to carrying out an actual simulation.

How a simulation works

The flowchart should have clarified the logic of the inventory problem. The simulation can now be carried out. Table 10.6 is a report of the demand distribution with associated random numbers. Table 10.7 is the lead-time distribution with associated random numbers. Table 10.8 shows the manual simulation of the inventory problem. In practice it would probably be done by computer. The manual version shows what is happening inside the black box.

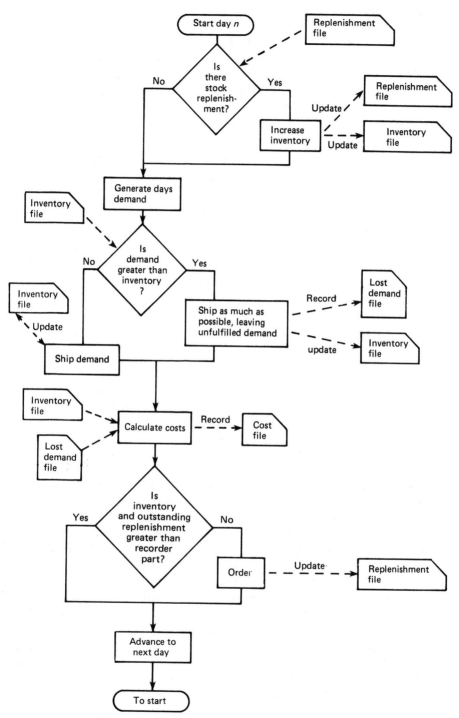

Figure 10.4   Flowchart of inventory problem

Table 10.6

Demand	Probability	Random numbers
1	0.10	00–09
2	0.15	10–24
3	0.25	25–49
4	0.25	50–74
5	0.15	75–89
6	0.10	90–99
	Total 1.00	

Table 10.7

Days	Probability	Associated random numbers
4	0.1	00–09
5	0.4	10–49
6	0.4	50–89
7	0.1	90–99
	Total 1.0	

The simulation is run for 40 days. There is nothing special about the choice of 40, but the choice should reflect a cycle which adequately covers the firm's business. For example, if there are quarterly seasonal variations in demand the period covered by the simulation should be at least 250 days to cover a working year (and, of course, the generation of daily demand should also reflect the seasonality).

On the first day the starting inventory is 30. This is a typical inventory level. In a short simulation this choice might significantly affect the results and, consequently, in some cases simulations are repeated using different starting conditions.

In a simulation the operator must define the policies to be tested. For this inventory problem the policies are defined by specifying the re-order quantity (the number of parts ordered when an order is placed) and the re-order point (the level of inventory which triggers off a new order). The first policy to be tested has been selected with these values at 40 and 20 respectively.

For the purposes of the example, two types of cost were specified: that of holding stock and that of unfulfilled demand. The stocking cost is £1 item/day and the cost of not being able to meet demand is estimated as £20/item.

For each day, column (A) shows the inventory at the start of that day. If the replenishment file shows a new order is to arrive that day, the order is added to the inventory. Demand for the day is generated using the Monte Carlo method (see table 10.9) and subtracted from the inventory to leave an end-of-day inventory in column (B). The cost of holding this inventory overnight is calculated in column (C). If inventory on hand is insufficient to fully meet demand, as much demand as possible is met (and inventory reduced to zero) and the remainder is recorded as lost demand. The cost of lost demand is calculated in column (D). Lastly, the end-of-day inventory is compared with the re-order point to see if a new order should be placed. If so, the quantity ordered, the lead-time before delivery (generated via Monte Carlo, see table 10.10), the date of arrival and the amount currently on order are recorded in the replenishment file. The simulation passes on to the next day.

Tables 10.9 and 10.10 show the generated demands and lead-times used in the simulation. The random numbers are taken from table 10.4. The first 4 rows were used for the demands, the 5th row for the lead-times.

Interpreting the output

At the end of the 40 days the total costs are calculated and recorded. Table 10.11 shows that for this policy the stockholding cost is £758 and the cost of the lost demand is £280, making a total of £1038. On 10% of days demand was not fully met. This can be checked from table 10.8.

Table 10.11 also shows the results of running the simulation for other inventory policies. The best appears to be with re-order quantity 30 and re-order point 20.

It is to be expected that a policy with a low re-order quantity is going to work out well. Such a policy would involve making a lot of re-orders each for the small quantity. Since the cost of the administration required for a re-order has not been included, this is not penalized. A more sophisticated simulation would include this cost.

Table 10.8 Inventory simulation

Inventory at start of Day 1 = 30
Re-order quantity = 40
Re-order point = 20
Stockholding cost = £1 item/day
Lost demand cost = £20/item

| | (A) | | | (B) | (C) | | (D) | | | | |
| | Inventory file | | | | Cost file | | | Replenishment file | | | |
Day	Start inventory	Receipts	Demand	End inventory	Stock cost	Lost demand	LD cost	Order placed	Lead time	Due in	On order
1	30		6	24	24			40	7	9	40
2	24		4	20	20						40
3	20		5	15	15						40
4	15		4	11	11						40
5	11		6	5	5						40
6	5		5	0	0						40
7	0		6	0	0	6	120				40
8	0		5	0	0	5	100				40
9	0	40	2	38	38						
10	38		4	34	34						
11	34		5	29	29						
12	29		6	23	23						
13	23		5	18	18			40	5	18	40
14	18		4	14	14						40
15	14		4	10	10						40

Row										
16	10		2	8						40
17	8		4	4						40
18	4	40	4	40						
19	40		1	39						
20	39		2	37						
21	37		2	35						
22	35		1	34			40			
23	34		3	31					32	
24	31		4	27				6		
25	27		5	22						
26	22		6	16						
27	16		6	10						40
28	10		3	7						40
29	7		4	3						40
30	3		4	0	1	20				40
31	0		2	0	2	40				40
32	0	40	4	36						40
33	36		5	31						
34	31		2	29						
35	29		2	27			40			
36	27		5	22				6	43	
37	22		5	17						40
38	17		2	15						40
39	15		1	14						40
40	14		1	13						40

Total = 758

Total = 280

Table 10.9 Simulated demand distribution

Day	Demand	Random numbers	Day	Demand	Random numbers	Day	Demand	Random numbers	Day	Demand	Random numbers
1	6	97	11	5	80	21	2	14	31	2	22
2	4	51	12	6	94	22	1	02	32	4	52
3	5	88	13	5	76	23	3	38	33	5	78
4	4	70	14	4	58	24	4	66	34	2	18
5	6	96	15	4	70	25	5	86	35	2	24
6	5	77	16	2	12	26	6	96	36	5	77
7	6	96	17	4	58	27	6	91	37	5	87
8	5	76	18	4	63	28	3	42	38	2	24
9	2	18	19	1	05	29	4	60	39	1	08
10	4	54	20	2	13	30	4	63	40	1	06

Table 10.10   Simulated lead times

Re-order	Lead-time	Random numbers	Re-order	Lead-time	Random numbers
1	7	96	6	5	10
2	5	44	7	4	08
3	6	82	8	6	68
4	6	84	9	7	90
5	6	70	10	6	76

The simulation has only tested the policies defined for it. It may be that policies such as (30,19), (30,21) etc., work better. There are very many combinations that could be tried to find a true optimum. The time and cost involved may not make it worth while and an approximately optimal solution may be the final result.

The results in table 10.11 are based on a particular set of random numbers, which generated the demand and lead-time variables. While the numbers were certainly random, purely by chance they may have been in some way unusual. The demands and lead-times would then not be representative of their distributions. For example, if the random numbers tended to be high, the demands would be high and the lead-times long. Repeating the simulation for the (40,20) policy but each time generating new sets of random numbers (and therefore of demands and lead-times) allows the policy to be evaluated on different data patterns. The results are in table 10.12. The policy gives total costs ranging between 787 and 1038. With such variation it would be dangerous to base a decision on a test over just one 40-day period. The simulation should be run many times and the average costs for each policy compared.

The number of simulation runs required in order to be reasonably satisfied with the final decision almost certainly means that even a simple simulation like the inventory problem needs to be put on the computer.

Table 10.13 shows the results of testing the same policies as in table 10.11. However, instead of just being from one 40-day simulation, the results are the averages of fifty 40-day simulations. Table 10.13 confirms the earlier results, and backs up the intuitive reasoning that a low re-order quantity policy will give the lowest costs of making orders.

Whilst several different random number streams should be used to evaluate a policy, it is important that the same streams should be

Table 10.11 Summary of simulation

Policy		Stockholding cost	Lost demand cost	Total	Percentage days out of stock
1. Re-order quantity	40	758	280	1038	10
Re-order point	20				
2. Re-order quantity	50	1067	220	1287	5
Re-order point	20				
3. Re-order quantity	30	627	220	847	5
Re-order point	20				
4. Re-order quantity	40	677	360	1037	12.5
Re-order point	15				
5. Re-order quantity	40	866	160	1026	5
Re-order point	25				

Table 10.12   Repeated simulations of policy 1

Trial	Stockholding cost	Lost demand cost	Total	Percentage days out of stock
1	758	280	1038	10
2	816	120	936	7.5
3	796	140	936	7.5
4	787	0	787	0

used to evaluate other policies. The reason is that the results from two policies should differ because the policies are different, not because they have been tested on different demand patterns. For example, in table 10.13 each policy was evaluated over 50 runs. The same 50 streams were used for each policy so that like is compared with like. Most computer programs cater for this by asking the user to specify the starting number. If the starts are the same, the streams will be identical; if different, the streams will differ. Each of the 50 runs had a different starting number, but the same 50 starting numbers were used for each policy.

A simulation does better than just measure the average costs of different policies. The risks can also be compared. Risk is indicated by the frequency of trials for which very high costs are recorded. Two policies may both have average costs over many runs of, say, 900 but if policy A had a range of 600–1200 for individual runs,

Table 10.13   Results averaged over fifty simulation runs

Policy		Stock cost	Lost demand cost	Total cost	Percentage days out of stock
1. Quantity Point	40 20	849	45	894	0.9
2. Quantity Point	50 20	1015	58	1073	1.1
3. Quantity Point	30 20	647	94	741	1.8
4. Quantity Point	40 15	718	206	924	3.7
5. Quantity Point	40 25	1018	2	1020	0.1

it would be regarded as more risky than policy B with a range of 850–950. In this case risk is also indicated by the percentage of days on which demand is lost. Lost demand, besides its financial implications, may also have important non-quantifiable disadvantages. The 'best' policy may then be decided by making a trade-off between average cost and percentage of days lost.

A relatively small computer program (about 100 lines) is sufficient to perform a simulation such as this. The key factors involved in using a simulation such as this are summarized in table 10.19 at the end of the chapter.

Risk analysis

The example just given was of a *physical* simulation. Other examples of physical simulations are the simulation of the arrivals and departures of ships at a port to determine the provision of facilities; and the simulation of a supermarket to determine the provision of checkout desks. The steel distribution example was also a physical simulation.

When a stochastic simulation uses the Monte Carlo technique to model cash flows it is called a *risk analysis*. For example, risk analysis could be used to compare ways of providing production facilities for some new product of a company. One of the stochastic variables might be the rate of growth of sales of the new product. In this case the simulation is run many times as if the capital decision were being taken many times over (compare the previous example where the simulation was run over many days). For each run a different growth rate is used (as selected via Monte Carlo) and a different cash flow (or internal rate of return) is calculated. Over many runs a distribution of cash flows is obtained. Not only is the average cash flow calculated, but also the risk (hence the name of the technique) of obtaining an unacceptably low profit from the project. A risk analysis does not differ from other stochastic simulations because any new technique is involved. It differs only in terms of its application (to financial problems).

Some controversy surrounds the simulation of present value distributions (but not cash flows or internal rates of return). The problem is that present values are based on discount rates. Since discount rates should be adjusted for risk, and since risk is also introduced via the distributions of input variables, a double allowance for risk is apparently made. Unfortunately, determining a discount rate is usually an approximate procedure. In practice, therefore, the simulation of present values is still useful, especially as the objective of a simulation is always to gain insight and not to provide a robotic decision-maker.

APPLICATION

A small town took its gas supply from a coal gas plant located in the town. The plant supplied gas only for the town and its immediate environment. It was, however, very old and had reached the point of being beyond repair. It was likely to cease functioning within 2 years. The Gas Board had three options:

1) Close the plant down, sell off the site and cease supplying gas to the town. This would mean that existing customers had to be compensated for expenses incurred in cutting off the supply and their need to purchase, for example, a new electric cooker.
2) Close the plant down, but build a liquid natural gas plant on the site. This would be virtually a pumping station. LNG would be brought to the town by tanker, converted to gas and pumped to customers just like the old coal-based gas. The plant is cheap, the LNG and cost of transport expensive.
3) Close the plant down and sell off the site, but extend a nearby (60 miles away) North Sea gas main and supply the town with North Sea gas. At the time of this decision most of the United Kingdom had just been covered with a network of North Sea gas mains which supplied most of the country but it had proved uneconomic to extend it to smaller places in rural areas. It was not originally intended to supply the small town in question with North Sea gas. Extending the main would be very expensive, but only a small pumping station would be needed in the town.

The Gas Board wanted to evaluate the three options by measuring their present values. A straightforward calculation of present value would, it was thought, not fully reflect the great uncertainties involved. The costs of extending the main, compensating cut-off customers and the rate of growth of demand, if, for example, North Sea gas should be available, were all variables which were not known with any certainty. It was decided to use risk analysis.

SIMULATION

The cash flows resulting from each of the three alternatives were simulated over a 25-year period. For each alternative 100 runs were made using different values for the 'uncertain' variables as generated by the Monte Carlo technique. The distributions of the 'uncertain'

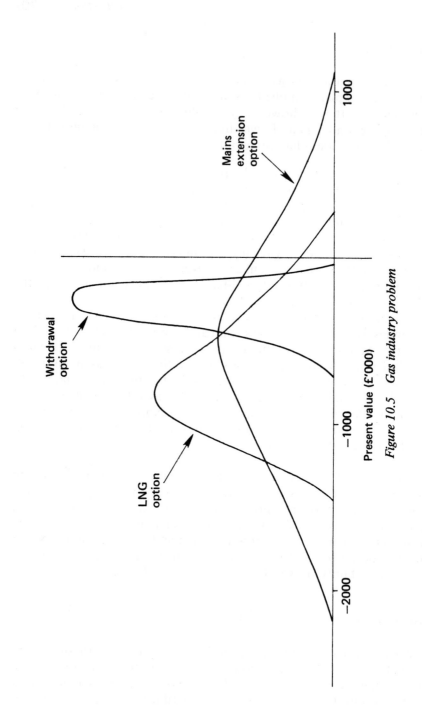

Figure 10.5   Gas industry problem

variables which were the basis of the Monte Carlo technique had to be obtained in a different way to those in the inventory problem. In that problem the demand and lead-time distributions were obtained from past records. The same was true (with the addition of some forecasting) for the demand distributions in the distribution of steel products example. In this example there were no past records and the distributions had to be assessed subjectively. For example, engineers had to use their judgment to assess, in distributional form, the capital cost of extending the North Sea gas main. There are standard techniques for doing this (see bibliography). The result was distributions of capital building costs, demand growths, operating costs, etc. which in the eyes of managers more accurately represented what was likely to happen than the single-point estimates which would otherwise have been used.

The output was, for each alternative, 100 calculations of present value represented as a distribution (see figure 10.5 and table 10.14). The decision as to which option to adopt was taken in the light of information about the likely spread of financial results, rather than just the average. In other words, the risk associated with an alternative was taken into account as well as the likely present value.

Table 10.14   Results of gas simulation (£'000)

Option	Average present value	Range of present values
Withdrawal	−193	−710 → −50
LNG	−692	−1400 → +250
Extension	−471	−2200 → +1200

Worked examples

1. W. E. TAPPER

This company has devised a new garden game based on cricket for up to 4 people. They plan to market it in a time for next summer. The sales manager has estimated the demand (table 10.15). How would this distribution be sampled for use in a simulation of the profitability of the new game? In particular what would be the demand associated with the numbers 79, 18 and 52 from a two-digit random number table?

Table 10.15

Units	Probability (%)
10,000–12,000	5
12,000–14,000	20
14,000–16,000	35
16,000–18,000	25
18,000–20,000	10
20,000–25,000	5

First associate two-digit numbers with the different demand levels in proportion to the probabilities. Second, let the midpoint of each category represent the whole category, e.g. the 10–12 class is represented by 11. This is an approximation, but since the original distribution was subjectively estimated, the approximation is reasonable in the context (table 10.16). A number in the range 00–04 is associated with a demand of 11,000 units, 05–24 with 13,000 units and so on. In particular 79, 18, and 52 will be associated with 17,000 units, 13,000 units and 15,000 units respectively.

Table 10.16

Midpoint ('000)	Units ('000)	Probabilities (%)	Numbers
11	10–12	5	00–04
13	12–14	20	05–24
15	14–16	35	25–59
17	16–18	25	60–84
19	18–20	10	85–94
22.5	20–25	5	95–99

2. ROYAL OVERSEAS BANK

A branch bank has eight serving counters, but not all of them are manned at any one time. The bank wishes to simulate its operations to determine how many should be manned between 9.30 and 12.00. During these hours the distributions of (a) the number of customer arrivals per minute and (b) the service time (the time a customer is actually dealing with a cashier) are both known, having been derived from a recent survey. A new customer will always join the shortest queue. The output of the simulation will consist of information on the percentage time cashiers are busy, average queue length, and average time a customer spends waiting.

Draw a flowchart for the simulation.

The simulation will proceed minute by minute between the hours of 9.30 and 12.00 (figure 10.6).

3. LONDON AND SOUTHERN DRUG CO.

The company has two new medical products at the development stage. Both require the same initial investment. Budget restrictions mean that it can launch either product but not both. A risk analysis of the cash flows over a 2-year period produces the results shown in table 10.17.

(a) For each product what is the probability of

    (i) a loss?
    (ii) a profit exceeding £300,000?

(b) Which product should be launched?

(a) If the distributions of profits are normal, the attributes of normal curves can be used to calculate the probabilities (table 10.18).

    For example, for product B the probability of making a loss is:

$$= P \text{ (profit is less than 0)}$$

$$= P \left(z \text{ value is less than } \frac{0 - 150}{150} = -1 \right)$$

Since 68% of a normal distribution lies between ± 1 standard

$$P \text{ (loss)} = 16\% \text{ (figure 10.7)}.$$

If the normality assumption cannot be made, then the simulation would have to print out the distributions as histograms, from which the probabilities can be measured directly. For example, the histogram for product B might be as shown in figure 10.8.

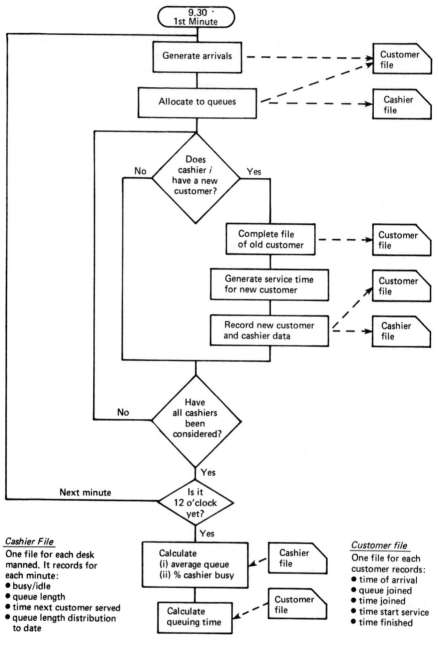

Figure 10.6

Table 10.17   Results from 100 iterations

	Average profit (£'000)	Standard deviation
Product A	120	40
Product B	150	150

Table 10.18   Probability

	Loss	Profit £300,000+
Product A	less than 1%	Virtually 0
Product B	16%	16%

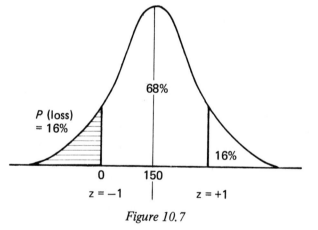

Figure 10.7

(b) Product B has a higher profit than A. If the company has many such investment opportunities then B would be preferable because on average it will produce the better return.

Product B is also more risky. It has a larger chance of a loss and of a high profit. If the company has few investment opportunities, and if the potential losses are large in comparison to its total capital, the company might prefer not to risk making a loss and select A. On the

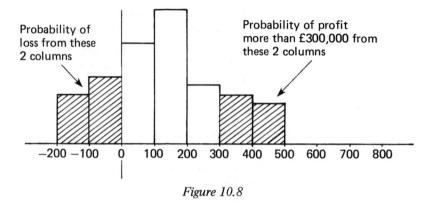

Figure 10.8

other hand the chance of a high profit might in itself be attractive in spite of the potential losses and, therefore, B might be preferable.

In short the choice between high average profit/high scatter and low average profit/low scatter depends upon the finances and risk attitudes of the company.

Final comments

Simulation is a much less restrictive aid to decision-making than the optimization techniques such as mathematical programming and decision analysis. The preconditions for being able to use the optimizers (e.g. linearity in LP) do not apply to simulation. This is one of the major reasons for its use. It has the further advantage that it is not as mysterious to non-quantitative managers as some of the mathematical techniques. Most people can understand what a simulation is by reference to analogies such as wind tunnels. As a result they are more likely to have confidence in simulation and to make use of it in their decision-making.

On the other hand there are some drawbacks in using simulation which may not be readily apparent. The major shortcomings are:

1   It can be an expensive technique. Simulation models tend to be large, which means they take a long time to develop and are expensive.
2   The data on which the simulation is based are often not readily available. The collection of the data, especially the subjective distributions, may take some time.
3   It relies on the creativity of the operator. The technique can

only indicate the best policy of those tried. If only poor policies are tried out, a poor decision will result.

4 Establishing the validity of the model is time-consuming. A rigorous series of tests may be required to determine the accuracy of the calculations and to throw up any logical errors in the methodology, especially when it is computer-based.

5 Since the nature of simulation (non-optimizing) requires many different policies to be tried out, it is inevitable that it will take time to get a reasonable answer from it.

It is rare for a manager to have to develop a simulation himself. His interest is more likely to be in the management of its development as with, for example, the construction of a corporate model. In this case he will need to be aware, in general, of what a simulation can do, and of its drawbacks.

On the other hand, he may well wish to use a simulation. For instance, he may wish to ask 'what if' questions of a corporate model. Some key factors in the use of a simulation are shown in table 10.19. They were discussed earlier in the context of the inventory example. Perhaps the most important of these factors is to adopt a structured approach to the simulation. If many policies are feasible then, if the simulation is computerized, it is easy to try out policies almost at random. A better approach is to plan a series of trial policies which are so structured that the range of options can be narrowed down until a nearly optimal one is obtained. Initially policies covering the whole spectrum of possibilities should be tried. This should indicate obviously unreasonable options, and suggest sections of the spectrum that need further investigation. These sections will be explored and a further narrowing down will be possible and so on. It is especially necessary to structure an approach in examples such as the inventory problem and the steel distribution problem where there are a wide range of possibilities.

Table 10.19   Key factors in using a simulation

---

1. Structured approach
2. Length of period covered
3. Number of runs
4. Same/different random number streams
5. Influence of starting conditions

In other cases, for example the gas industry problem, the options are predefined and structuring becomes less vital.

Further example

SOUTH LONDON OMNIBUS WORKSHOP

The Workshop is a local government organization responsible, for amongst other things, servicing the vehicles of the public-owned local bus company. By law, each week every bus has to be inspected. The inspection takes 30 minutes, after which, if the bus has no faults, it returns to duty. If the bus has minor faults they are repaired there and then. If the bus has a major fault it is taken elsewhere for repairs. The time spent by a bus at the inspection facilities can be as little as 30 minutes (if it has no faults or has major faults). Should minor repairs be required the time spent can be up to 4 hours. The distribution of time spent at the facilities has been constructed from historical records and is shown in table 10.20. The time is to the nearest 30 minutes because this is what is shown on the inspection records. Buses are taken to the inspection centre when schedules allow. The arrival rate distribution has been derived from past records and is shown in table 10.21. The inspection facilities comprise three identical and adjacent units, each of which deals with one bus at a time.

Because of reductions in public spending, and because it is felt

Table 10.20

Time spent (minutes)	Probability (%)
30	65
60	5
90	3
120	4
150	6
180	9
210	5
240	3
	100

Table 10.21

Number of buses arriving in 30-minute period	Probability (%)
0	25
1	35
2	20
3	10
4	8
5	2
	100

that the units are unused for a lot of each day, it is planned to close one of the units. This has caused a great many complaints from unions and local politicians who say that the reduction will lower safety standards and be uneconomic (because of the greater time buses will spend waiting for an inspection).

The Chief Traffic Engineer has decided to simulate operations at the inspection facilities to produce facts and figures concerning the effect of having two units compared to three.

(1) How should the distributions be handled in the simulation?
(2) Produce a flowchart for the simulation.
(3) What factors are relevant to the argument taking place, i.e. what should the output of the simulation be?

ANSWERS

(1) Monte Carlo technique.

(2) See figure 10.9.

(3) Economically the argument involves the trade-off between the time buses/crews are kept waiting for inspection against the time the inspection units may be idle waiting for buses to arrive. Safety might be affected if the queues of waiting buses were long so that the inspection mechanics hurried their work and made mistakes. The

Time Period = 30 minutes

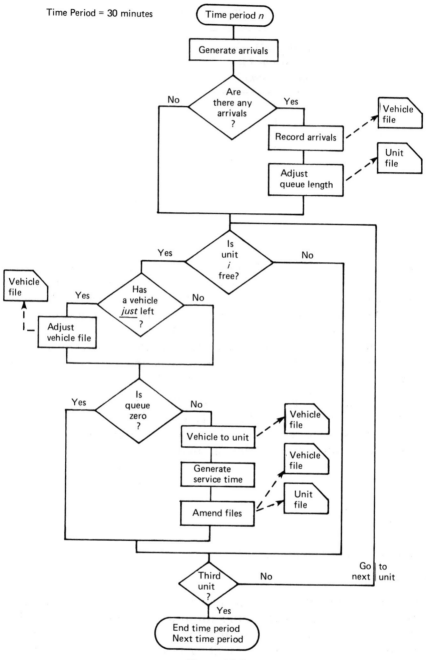

Figure 10.9

relevant factors would seem to be:

Operational: Average time/week units are unused
              Average time a bus is queuing for an inspection
              Average number of buses in a queue
              Maximum queue length

Financial: Cost of crews having to wait
              Cost of buses having to wait
              Cost of units being idle
              Saving in running costs from having two units only.

Appendices

Appendix A: Probability

All future events are uncertain to some degree. That the present government will still be in power in the UK in a month's time (assuming this is not an election month) is likely, but far from certain; that a communist government will be in power in a month's time is highly unlikely, but not impossible. Probability theory enables the difference in the uncertainty of events to be made more precise by measuring their likelihood on a scale. The scale is shown in figure A.1. At one extreme, impossible events (e.g. that you could swim the Atlantic) have probability zero. At the other extreme, completely certain events (that you will one day die) have probability one. In between are placed all the neither certain nor impossible events according to their likelihood. For instance, the probability of obtaining a head on one spin of an unbiased coin is ½; the probability of one particular ticket winning a raffle in which there are 100 tickets in total is 0.01.

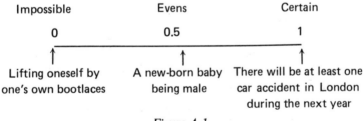

Figure A.1

As a shorthand notation 'the probability of an event A is 0.6' is written in this way:

$$P(A) = 0.6$$

Measurement of probability

There are three methods of calculating a probability. The methods are not alternatives since for certain events only one particular method of measurement may be possible. However, they do provide different conceptual ways of viewing probability. This should become clear as the methods are described.

(a) '*A PRIORI*' APPROACH

In this method the probability of an event is calculated by a process of logic. No experiment or judgment is required. Probabilities involving coins, dice, playing cards, etc., can fall into this category. For example, the probability of a coin landing 'heads' can be calculated by noting that the coin has 2 sides, both of which are equally likely to fall upwards (pedants note: assume it will not come to rest on its rim). Since the coin must fall with one side upwards, the two events must share equally the total probability of 1.0. Therefore

$$P(\text{Heads}) = 0.5$$
$$P(\text{Tails}) = 0.5$$

(b) 'RELATIVE FREQUENCY' APPROACH

When the event has been or can be repeated a large number of times, its probability can be measured from the formula

$$P(\text{event}) = \frac{\text{No. of times event occurs}}{\text{No. of trials}}$$

For example, to estimate the probability of rain on a given day in September in London, look at the last 10 years' records to find that it rained on 57 days. Then

$$P(\text{rain}) = \frac{\text{No. of days rain recorded}}{\text{Total no. of days} (= 10 \times 30)}$$
$$= \frac{57}{300}$$
$$= 0.19$$

(c) SUBJECTIVE APPROACH

A certain group of statisticians (Bayesians) would argue that the degree of belief that an individual has about a particular event may be expressed as a probability. Bayesian statisticians argue that in certain circumstances a person's subjective assessment of a probability can and should be used. The traditional view, held by classical statisticians, is that only objective probability assessments are permissible. Specific areas and techniques that use subjective probabilities will be described later. At this stage it is important to know that probabilities can be assessed subjectively but that there is discussion amongst statisticians as to the validity of doing so.

As an example of the subjective approach, let the event be the achievement of political unity in Europe by the year AD 2000. There is no way that either of the first two approaches could be employed to calculate this probability. However, an individual can express his own feelings on the likelihood of this event by comparing it with an event of known probability, e.g. is it more or less likely than obtaining a head on the spin of a coin? After a long process of comparison and checking, the result might be:

$$P(\text{political unity in Europe by AD 2000}) = 0.10$$

The process of accurately assessing a subjective probability is a field of study in its own right and should not be regarded as pure guesswork.

The three methods of determining probabilities have been presented here as an introduction and the approach has not been rigorous. Once probabilities have been calculated by whatever method, they are treated in exactly the same way.

Examples

(1) What is the probability of throwing a 6 with one throw of a die?

A priori approach – there are six possible outcomes: 1, 2, 3, 4, 5 or 6 showing.

All outcomes are equally likely, therefore

$$P(\text{throwing a 6}) = \frac{1}{6}$$

(2) What is the probability of the channel tunnel being completed by AD 2025?

The subjective approach is the only one possible, since no logical chain of thought could lead one to an answer and there are no past observations. My assessment is a small one; around 0.02.

(3) How would you calculate the probability of obtaining a head on one spin of a biased coin?

The *a priori* approach may be possible if one had information on the aerodynamical behaviour of the coin. A more realistic method would be to conduct several trial spins and count the number of times a head appeared:

$$P(\text{obtaining a head}) = \frac{\text{No. of observed heads}}{\text{No. of trial spins}}$$

Appendix B: Graphs

Graphs are the major pictorial method of representing numbers. Changes in, for instance, sales figures or financial measures over time are immediately apparent; the relationship between variables, say supply and demand, are quickly evident. Graphs are used in most areas of management.

The essence of graphical representation is the location of a point on a piece of paper by specifying its co-ordinates. As an example, consider a town map. An entry in the index might read:

Acacia Avenue P7, F2

Turning to page 7, the map is divided into rectangles (see figure B.1). To find Acacia Avenue, look along row F, and down column 2. Where row and column meet is rectangle F2, somewhere within which those with acute eyesight will be able to find the avenue in question.

More generally. any point on a graph is located by two co-ordinates, which are the horizontal and vertical distances of the point from a

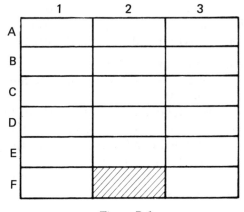

Figure B.1

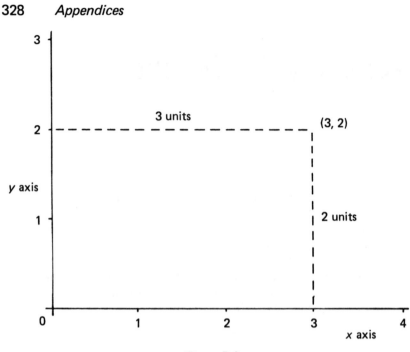

Figure B.2

fixed point called the origin. Figure B.2 illustrates this for the point (3,2). This point is 3 units from the origin in a horizontal direction (the horizontal scale is usually called the x-axis); it is 2 units from the origin in a vertical direction (the vertical scale is usually called the y axis). The first co-ordinate is referred to as the x-value and the second co-ordinate as the y value. By convention, the order is always the same. Logically developing the above, the following facts emerge:

- The origin has co-ordinates (0,0).
- Anywhere along the y-axis, the x-value is 0. Similarly, along the x-axis, the y-value is 0.
- The axes can be extended on the other side of the origin. In this case, either or both co-ordinates take on negative values. When the scales are extended in all four directions, the paper is divided into 4 quadrants (see figure B.3)
- There is no restriction to whole units, therefore *any* point on the paper has a representation.

Figure B.4 gives some examples of point representation.

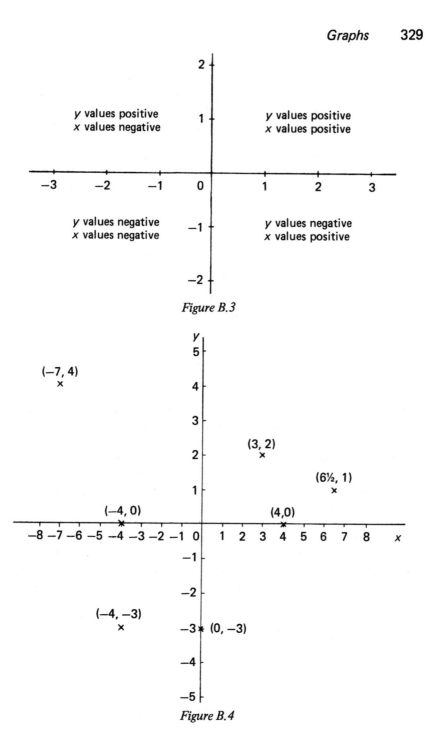

y values positive
x values negative

y values positive
x values positive

y values negative
x values negative

y values negative
x values positive

Figure B.3

(−7, 4)
×

(3, 2)
×

(6½, 1)
×

(−4, 0)
×

(4,0)
×

(−4, −3)
×

(0, −3)

Figure B.4

Graphical representation is not limited to the location of points. Relationships between two variables can be illustrated on a graph. Consider the simple example of the direct profit when a product is sold. For a given price and cost of the product, the direct profit will vary accordingly to the numbers sold.

$$\text{Direct profit} = (\text{Price} - \text{Cost}) \times \text{Volume}$$

This can be written more concisely if letters stand in place of the words. Suppose x is the variable number of products sold, y is the variable direct profit, p the set price and q the set cost per unit. Then

$$y = (p - q)x \qquad (1)$$

The equation is given a label (1) so that it can be referred to later. Note that multiplication can be shown in several different ways. For example, Price (p) times Volume (x) can be written as:

$$p.x$$

$$(p)(x)$$

$$px$$

The multiplication sign ($\times$) used in arithmetic tends not to be used in algebra because of possible confusion with the use of x as a variable.

The use of symbols to represent numbers is the definition of algebra. It is not done to confuse but to simplify. The symbols (as opposed to verbal labels, i.e. 'y' instead of 'Direct profit') shorten the description of a complex relationship; the symbols (as opposed to the numbers, i.e. y instead of 2.1, 3.7, etc.) allow the general properties of the variables to be investigated.

(1) is an equation. Since price and cost are fixed in this example, p and q are constants. Depending upon the quantities sold, x and y may take on any of a range of values, therefore they are variables. Once x is known, y is automatically determined. y is said to be a function of x.

If the values of the constants are known, say $p = 5$, $q = 3$, then the equation becomes:

$$y = 2x \qquad (2)$$

A graph can now be made of this function. The graph is the set of all points satisfying (2), i.e. all the points for which (2) is true. By looking at some of the points the shape of the graph can be seen:

when $x = 0$, $y = 2$ times $0 = 0$

when $x = 1$, $y = 2$ times $1 = 2$

when $x = 2$, $y = 2$ times $2 = 4$, etc.

Therefore, points (0,0), (1,2), (2,4), etc., all lie on this function. Joining together a sample of such points shows the shape of this graph. Figure B.5 shows the graph of the function $y = 2x$.

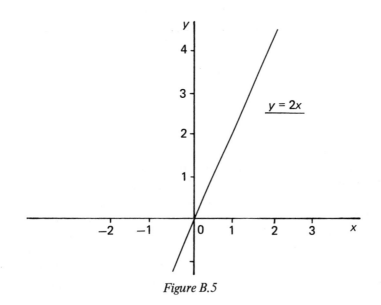

Figure B.5

This function is called a linear function since it is a straight line (or, more mathematically, it is linear because y and x are not raised to powers. There are no squared, cubed, logarithmic, etc., terms). More complicated functions can be graphed, for instance, ones including x^2. $y = x^2 - 2$ (figure B.6) might represent the relationship between a firm's profit, y, and output, x (but only for x greater than 0). Profit is at first negative; a breakeven point is reached at $x = 1.4$ (approx.); thereafter profit increases rapidly as economies of scale come into play. At the present it has not been demonstrated that the equation $(y = x^2 - 2)$ could, in certain circumstances, represent the relationship between profit and output.

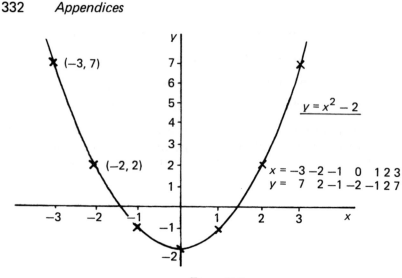

Figure B.6

Functions are put into graphical form by plotting a selection of points and then drawing the curve that goes through.

In figure B.6, $x = -3$ might be the starting point. The corresponding y value is 7 ($x = -3$) is put into $y = x^2 - 2$ to give the y value). The point $(-3,7)$ therefore lies on the graph of the function. Trying $x = -2$ gives $y = 2$. The point $(-2,2)$ therefore lies on the graph. Continue plotting points until there are sufficient to indicate the shape of the graph. In this case taking all whole-number values of x between -3 and $+3$ gives sufficient points for the shape to become obvious.

$$x = -3 \quad -2 \quad -1 \quad 0 \quad 1 \quad 2 \quad 3$$
$$y = 7 \quad 2 \quad -1 \quad -2 \quad -1 \quad 2 \quad 7$$

The choice of these 7 points is arbitrary. A selection of points that are able to demonstrate the shape are chosen.

In figure B.5 only 2 points need be plotted since a straight line is defined completely by any 2 points lying on it.

Appendix C: Linear functions

Linear functions are of great importance in management studies. Not only do they describe in their own right many relationships, but also, because of the simplicity and ease of use, more complex relationships can sometimes be approximated by the linear equation. For instance, it might be possible for the total cost of processing particular benefit payments to be expressed as a linear function of the number of applications. The equation could then be used, along with others of a similar nature, in compiling a budget for the department concerned.

A linear function of x is one in which only constants and multiples of x appear. There are no terms such as x^2, x^5, $1/x$, etc. Because of this, if y is a linear function of x, then it can be deduced that the equation must have the form

$$y = a + bx \qquad\qquad (3)$$

where a and b are constants.

(3) is the general form of a linear equation; a and b are merely labels; any other letters would be just as satisfactory. In the benefit payments example, a is the fixed overhead, b the variable cost per application, x is the number of applications. $a + bx$ is then the total cost $(= y)$.

An alternative definition is that the graphical representation of a linear function must be a straight line, for at all points a change of 1 unit in x gives rise to a change of b units in y. The following are all straight lines:

$$y = 2x + 1$$

$$y = 3x$$

$$y = 4 - 2x$$

Interpretation of *a* and *b*

Figure C.1 is the graph of $y = 2x + 1$. It was obtained, as previously, by plotting two of the points. In this example $b = 2$ and $a = 1$.

The value of a is the intercept of the line. It is the point on the y-axis where the line crosses. This can be seen either from the graph or by putting $x = 0$ in the equation.

b is the slope of the line. Alternatively, the slope is referred to as the gradient. In either case, what is meant is the usual idea of gradient – the ratio between the distance moved vertically and the distance moved horizontally.

A few further points are worthy of note. First, it is possible for b to be negative. If this is the case the line leans in a backward direction, since as x increases, y decreases. This is illustrated in figure C.2.

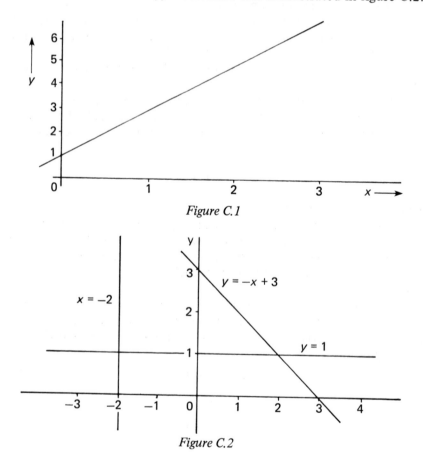

Figure C.1

Figure C.2

Secondly, it is possible for b to take the value 0. The equation of the line is then $y =$ constant, and the line is parallel to the x-axis. Similarly, the line $x =$ constant is parallel to the y-axis and the slope can be regarded as being infinite. The last two lines are examples of constant functions and are also shown in figure C.2.

Appendix D: Exponential functions

Exponential functions are important because of the way they express growth or decay (e.g. the increase or decrease in a variable over a period of time). Usually they represent a 'natural' pattern of growth or decay over time.

Think of this in terms of the sales of a new product. y is the level of sales and x is time. Under a linear growth function each month shows the same constant increase in the number (of bottles, packets, boxes, etc.) sold. Under an exponential growth function each month shows a different sales increase which is a constant proportion of the sales level at the start of the month. Obviously, there are many occasions when exponential growth is the more realistic assumption.

For the linear function $y = a + bx$ if x increases by 1 unit then y increases by b units. It does not matter whether x increases from 10 to 11 or from 100 to 101, y still increases by b units. For an exponential function, however, if x increases by 1 unit, then y increases by some constant percentage of itself. In this case it does matter whether x increases from 10 to 11 or from 100 to 101. y increases by a constant percentage of its value at $x = 10$ or $x = 100$. If the y values are different at $x = 10$ and $x = 100$ then so will be the amount by which y increases.

Exponential functions model decreases as well as increases.

Before studying this function in detail it is first necessary to look at the properties of exponents and logarithms, on which exponential functions are based.

Exponents

Consider an expression of the form a^x. The base is a and the exponent is x. If x is a whole number then the expression has an obvious meaning (e.g. $a^2 = a \times a$, $a^3 = a \times a \times a$, $3^4 = 81$, etc.) It also has meaning for values of x that are not whole numbers. To see what this meaning is, it is necessary to look at the rules for working with exponents.

Multiplication   the rule is:

$$a^x \times a^y = a^{x+y}$$

e.g. $a^2 \times a^3 = a^5$

$$b^2 \times b = b^3$$

It can be seen that this makes good sense if one substitutes whole numbers for a, x and y. For instance:

$$2^2 \times 2^3$$

$$= 4 \times 8$$

$$= 32$$

$$= 2^5$$

Note that the exponents can only be added if the bases are the same, e.g.:

$a^3 \times b^2$ cannot be simplified

Division   The rule is similar to that for multiplication:

$$a^x \div a^y = a^{x-y}$$

e.g. $x^2 \div x = x^{2-1} = x^1 = x$

Again the reasonableness of the rule is confirmed by resorting to a specific numerical example.

Raising to a power   The rule is:

$$(a^x)^y = a^{xy}$$

e.g. $(a^2)^2 = a^4$

Several points of detail follow from the rules:

(1) $a^0 = 1$ (since $1 = a^3/a^3 = a^{3-3} = a^0$)

(2) $a^{-x} = 1/a^x$ (since $1/a^x = a^0/a^x = a^{0-x} = a^{-x}$

(3) $a^{\frac{1}{2}} = \sqrt{a}$, $a^{\frac{1}{3}} = \sqrt[3]{a}$ (since $a^{\frac{1}{2}} \times a^{\frac{1}{2}} = a^{\frac{1}{2}+\frac{1}{2}} = a^1 = a$)

EXAMPLES

(1) Evaluate $(2^2)^3$

$$= 2^6$$

$$\underline{= 64}$$

(2) Evaluate $27^{\frac{4}{3}}$

$$27^{\frac{4}{3}} = (27^{\frac{1}{3}})^4$$

$$= (^3\sqrt{27})^4$$

$$\underline{= 3^4}$$

Logarithms

In pursuing the objective of understanding exponential functions, it is also helpful to look at logarithms. At school logarithms are used for multiplying and dividing large numbers, but this is not the purpose here. A logarithm is simply an exponent. For example, if

$$y = a^x$$

then x is said to be the logarithm of y to the base a. This is written as $\log_a y = x$

EXAMPLES:

(1) $1000 = 10^3$ and therefore the logarithm of 1000 to the base 10 is 3, i.e. $3 = \log_{10} 1000$

Logarithms to the base 10 are known as common logarithms.

(2) $8 = 2^3$ and therefore the logarithm of 8 to the base 2 is 3, i.e. $3 = \log_2 8$

Logarithms to the base 2 are binary logarithms.

(3) e is a component frequently found in mathematics (just as π is a constant found frequently in mathematics). e has the value 2.718 approx. Logarithms to the base e are called natural logarithms and are written ln;

i.e. $x = \ln y$ means $x = \log_e y$

e has other properties which make it of interest in mathematics.

The rules for manipulation of logarithms follow from the rules for exponents:

Addition: $\log_a x + \log_a y = \log_a xy$

Subtraction: $\log_a x - \log_a y = \log_a (x/y)$

Multiplication by a constant: $c\log_a y = \log_a y^c$

Exponential functions

The general form of the exponential function is

$$y = ka^{cx}$$

Putting $x = 0$ into the equation reveals that k is the intercept. The exponent constant c is more interesting. If it is positive, then the relationship is one of growth, that is as x increases, y also increases; if it is negative, then the relationship is one of decay, that is, as x increases, y decreases. Figure D.1 gives examples of exponential functions of growth and decay. Recall the property of the exponential function which makes it applicable in certain circumstances. If

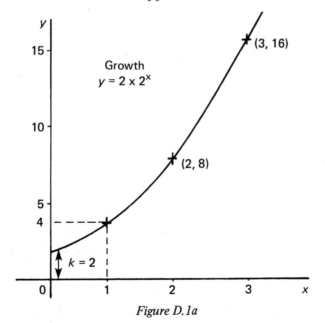

Figure D.1a

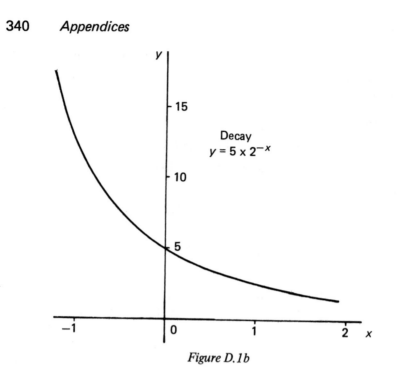

Figure D.1b

the relationship is one of growth then if x increases by a constant amount, y increases by a constant percentage of itself, e.g. in Figure D.1(a) 3 points are (1,4), (2,8), (3,16); in each case x increases by 1 and y increases by 100%. Similarly, for decay functions, if x increases by a constant amount, then y decreases by a constant percentage of itself.

EXAMPLE

The occurrence of a particular illness has been at a constant level of 100 new cases/year for some time. A new treatment has been developed and it is thought that the number of new cases will now decrease at a constant percentage of 25 per annum. What is the exponential function that describes this pattern? A graph of the function will be as in figure D.2. The general form of the exponential function is:

$$y = ka^{cx}$$

The base to work with can be chosen. The base 10 is selected here,

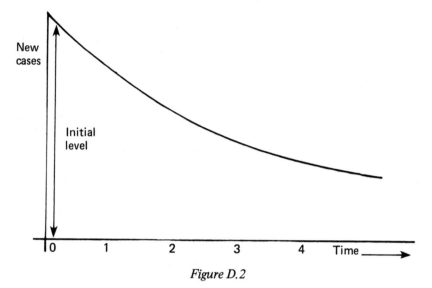

Figure D.2

but any other base would do as well.

$$y = k10^{cx}$$

Initially (at the beginning of the period, when $x = 0$) the level of new cases is 100, therefore

$$100 = k10^0 = k$$

So $$y = 100.(10^{cx})$$

After the first year, the new cases have dropped by 25% to 75.

$$75 = 100(10^c)$$

$$10^c = 0.75$$

At this point, tables, or a calculator have to be consulted. Logarithmic tables show the logarithm of a number, given the number. In this case the number is 0.75 and we need to know its logarithm c. The tables reveal that

$$c = -0.125$$

The exponential function relating y and x is therefore:

$$y = \underline{100.10^{-0.125x}}$$

Figure D.2 is a plot of this function. In this situation estimates were being made for the future based on judgments of a subjective nature. The judgments were: (1) an exponential function was correct, (2) the rate of decrease was 25%. The equation gives anticipated numbers of new cases at future points in time.

In other circumstances the problem might have been the other way round. The historical record of numbers of cases may have been available and the task may have been to find the function that best described what had happened. After the data were plotted on a graph there would be two problems. First, it would have to be judged whether the shape of the pattern of points was most like a straight line or an exponential curve or one of the many other types of functions that could be used. This judgment would be based on knowledge of the general shapes of these functions. The second problem would be that, even if a straight line, say, were judged to be the most likely shape, not all the points would fall precisely on a straight line (see figure D.3). The second problem is, then, to decide which straight line best fits the data points. A statistical technique, called regression analysis, is the usual way of dealing with this problem.

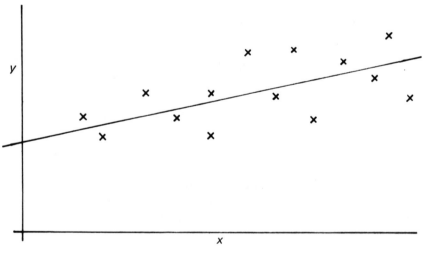

Figure D.3

Relationship between linear and exponential functions

The exponential function is:

$$y = ka^{cx}$$

Take logarithms to the base a on either side

$$\log_a y = \log_a (ka^{cx})$$

Apply two of the logarithmic rules (see page 339) to the right-hand side,

$$\log_a y = \log_a k + \log_a (a^{cx}) \quad \text{(Addition)}$$

$$= \log_a k + cx \log_a a \quad \text{(Multiplication by a constant)}$$

Since $\log_a a = 1$,

$$\log_a y = \text{constant} + cx$$

Therefore the relationship between $\log_a y$ and x is a linear one. (Compare this last equation with the equation of a straight line.) If y and x are related by an exponential function, then $\log_a y$ and x are related by a linear function. In other words, by means of a transformation (taking logarithms of both sides) an exponential function can be treated as a linear one. Such a transformation facilitates the use of regression analysis.

Appendix E: Simultaneous equations

Relationships between variables can be described by functions. In particular, linear equations can represent the linear relationship between two variables. There are, however, situations which are described by several equations.

For example, in microeconomics the price and production level of a product can be written as two equations. First, a relationship between price and quantity will show the amount consumers will demand at a given price level. Second, a further relationship between price and quantity will show the quantity a supplier will be willing to supply at a given price level. Economic theory says that there is an equilibrium point at which a single price and quantity will satisfy both equations (the demand curve and the supply curve).

This is the type of problem to be investigated. How can values for variables be found which satisfy simultaneously more than one equation? Only linear equations will be considered.

Suppose there are two equations in the two variables x and y as below:

$$3x + 2y = 18 \qquad\qquad (4)$$

$$x + 4y = 16 \qquad\qquad (5)$$

What is the value of x and the value of y which satisfies both these equations? Do such values exist at all? Or are there several of them? For example, can a single price and a single quantity be found that is correct for both the supply and demand equations?

Types of solution

If the equations are plotted on a graph, it is easier to understand the situation. The lines are best plotted by determining the points at which they cross the two axes (figure E.1).

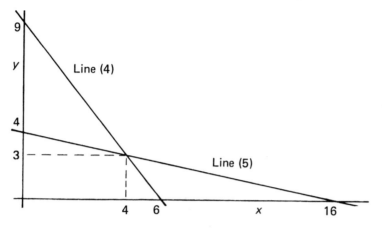

Figure E.1

For line (4):   when $x = 0$, $y = 9$

when $y = 0$, $x = 6$

For line (5):   when $x = 0$, $y = 4$

when $y = 0$, $x = 16$

The values of x and y which satisfy both equations are found from the point of intersection of the lines. Since this point is on both lines, the x and y values here must satisfy both equations. From the graph these values can be read: $y = 3$, $x = 4$. That these values do fit both equations can be checked by substituting $y = 3$, $x = 4$ into the equations of the lines.

In this example there is one and only one answer to the problem. There is said to the a unique solution to the problem.

On the other hand, suppose the 2 equations are as below:

$$2x + 3y = 12$$

$$2x + 3y = 24$$

In this case the two equations are inconsistent. The left-hand side of both equations is the same, $2x + 3y$. It is not possible for this to equal simultaneously 12 and 24. They can be plotted as in figure E.2. The two lines are parallel and have therefore no point of intersection and no common solution.

There is one other possibility that can arise. Suppose the equations

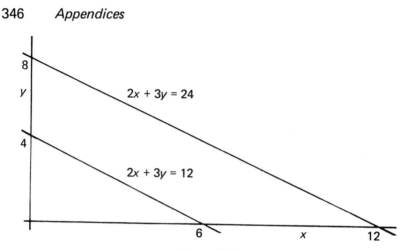

Figure E.2

are of the form:

$$x + 3y = 15$$
$$4x + 12y = 60$$

If these lines are plotted, it is found that they are exactly the same line. The second line is 4 times the first. Cancelling through by 4 gives the same equation. Any point lying anywhere along the two coincident lines satisfies both equations. There are an infinite number of solutions and the equations are said to be dependent.

To recap, solving 2 linear equations in 2 variables means finding a point whose co-ordinates satisfy both equations. One of three outcomes is possible:

- there is one common point, i.e. a unique solution;
- there are an infinite number of solutions, i.e. the equations are dependent;
- there is no solution, i.e. the equations are inconsistent.

Algebraic solution

In the preceding example the solutions, if any, to the equations were found graphically. The solutions can also be found by more mathematical means. The method works as shown in table E.1.

Table E.1

Method	Example
	$3x + 2y = 18$ $x + 4y = 16$
1. Multiply one or both the equations by appropriate numbers so that the coefficients of x (or y) are the same in both equations. (A coefficient in this context is a constant by which a variable is multiplied, e.g. for $2x$, 2 is the coefficient of x.)	1. Multiply the second equation by 3. Leave the first equation. The coefficient of x is 3 in both equations: $3x + 2y = 18$ $3x + 12y = 48$
2. Add or subtract the equations so that x (or y) disappears leaving one equation in y (or x).	2. Subtract the first from the second: $3x + 12y = 48$ $3x + 2y = 18$ $\overline{10y = 30}$
3. The remaining equation gives the value of y (or x).	3. $\quad 10y = 30$ $\quad\quad y = 3$
4. Substitute this answer in either of the original equations to provide the other half of the solution.	4. Put $y = 3$ in: $3x + 2y = 18$ $3x + 6 = 18$ $3x = 12$ $x = 4$
	The answer is $y = 3, x = 4$

Example

Solve the two simultaneous equations:

$$5x + 2y = 17 \qquad (6)$$
$$2x - 3y = 3 \qquad (7)$$

Multiply (6) by 3, (7) by 2, so that the coefficients of y are the same in both equations. ((6) could just as well have been multiplied by 2 and (7) by 5 and then x eliminated.)

$$15x + 6y = 51$$
$$4x - 6y = 6$$

Add the two equations to eliminate y,

$$19x = 57$$

$$x = 3$$

Substitute $x = 3$ in (7) to find the y value,

$$6 - 3y = 3$$

$$3y = 3$$

$$y = 1$$

The solution is $y = 1$, $x = 3$.

Further reading

Part I

The methods of data communication are based on the work of Professor Ehrenberg:

A. S. C. Ehrenberg (1982) *A Primer in Data Reduction*, Wiley & Sons.

The summary measures of chapter 3 are covered in

D. R. Anderson, D. J. Sweeney and T. A. Williams (1981) *Introduction to Statistics*, St Paul, Minnesota, West; and

J. E. Freund and F. R. Williams (1982) *Elementary Business Statistics*, 4th ed., Englewood Cliffs, N.J., Prentice-Hall.

Freund and Williams, *op. cit.* also has a separate chapter on indices.

Part II

There are many textbooks covering the statistical methods of chapters 4 and 5. Three of the best are

D. R. Anderson, D. J. Sweeney and T. A. Williams, *op. cit.*

J. E. Freund and F. R. Williams, *op. cit.*; and

P. G. Moore (1979) *The Principles of Statistical Techniques*, Cambridge, Cambridge University Press. This book shows the derivation of the formulae used in chapter 4.

Part III

The following three books give good basic explanations of regression and correlation:

D. R. Anderson, D. J. Sweeney and T. A. Williams, *op. cit.*

J. E. Freund and F. R. Williams, *op. cit.*; and

Tom Cass (1973) *Statistical Methods in Management*, London, Cassells.

To delve more deeply in regression and correlation, consult

J. Johnston (1972) *Econometric Methods*, New York, McGraw-Hill.

For a comprehensive coverage of statistical forecasting methods:

Spyros Makridakis and Steven C. Wheelwright (1978) *Forecasting Methods and Applications*, New York, Wiley & Sons.

For advanced forecasting methods together with applications:

Gwilym M. Jenkins (1979) *Practical Experiences with Modelling and Forecasting Time Series*, Jersey, Gwilym Jenkins and Partners.

Part IV

General books covering decision-making in some depth:

D. R. Anderson, D. J. Sweeney and T. A. Williams (1979) *An Introduction to Management Science*, St Paul, Minnesota, West.

H. A. Taha (1982) *Operations Research: An Introduction*, London, Macmillan.

S. C. Littlechild (1977) *Operational Research for Managers*, Deddington, Philip Allan.

For applications of decision analysis:

P. G. Moore, H. Thomas, D. W. Bunn and J. M. Hampton (1976) *Case Studies in Decision Analysis*, Harmondsworth, Penguin Modern Management Series.

For a detailed study of decision analysis:

P. G. Moore and H. Thomas (1976) *Anatomy of Decisions*, Harmondsworth, Penguin Modern Management Series.

For advanced mathematical programming:

H. M. Wagner (1970) *Principles of Management Science*, Prentice Hall.

For probability, decision analysis and simulation:

Charles A. Holloway (1979) *Decision Making under Uncertainty*, Englewood Cliffs, N.J., Prentice-Hall.

Glossary

The *absolute value* of a number is its positive value ignoring the sign. For instance, the absolute value of -3 is $+3$.

Arithmetic mean $(\bar{x})$ is a measure of location defined by:

$$\bar{x} = \Sigma x/n.$$

The *base year* in an indexed time series is the year when the index equals 100.

Bayes Theorem is the means of revising probabilities in the light of extra information. The original probability of an event (prior probability) will change if extra information is available. The new probability (posterior probability) can be calculated from Bayes Theorem.

The *binomial distribution* is a standard distribution derived from the following situation: A population comprises elements of 2 types. A random sample is taken from the population. The binomial distribution gives the probabilities of obtaining different numbers of elements of the 2 types in the sample.

Causal modelling is that group of forecasting methods based on relating (usually by regression) the variable to be forecast with other variables thought to cause changes in it.

The *Central Limit Theorem* states that as the sample size increases the sampling distribution of the mean becomes more like a normal distribution whatever the nature of the underlying distribution.

Confidence levels are measures of the level of certainty attached to an estimate or significance test. A conclusion based on inference can never be drawn with 100% certainty. For example a 95% confidence level means that a conclusion would be expected to be correct 95 times out of 100 if similar situations arose.

Constraints in LP are algebraic expressions of the limitations on resource availability.

A *continuous distribution* is one for which the variable can take on all values. There are no gaps within the range of values it takes. It is measured on a continuous scale.

Correlation is the determination of the strength of the relationship between one variable and one or more other variables. The *correlation coefficient* is a measure of this strength.

Critical values mark the accept/reject boundaries in significance tests. A sample statistic on one side of a critical value implies that the hypothesis should be rejected; on the other side, it should be accepted.

Cross-sectional data are data collected at one point in time.

Curvilinear regression deals with curved relationships. It has squared cubed etc. variables on its right-hand side and treats them as if they were separate linear variables.

Cycles are regularly repeating patterns of upward and downward movements in a data series. They have a length (the time they take to repeat) of less than 1 year.

Decision analysis is a technique for solving decision problems that are characterized by comprising a series of subdecisions and chance events spread over time.

Decision trees are the structures which represent diagramatically a logical sequence of decisions and chance events. The points in the tree at which decisions are taken or chance events occur are called *nodes*. The lines linking the nodes are called *branches*.

Decision variables are variables the setting of which at some level constitutes taking a decision. Typically in LP they are production levels.

Delphi is a qualitative forecasting method based on bringing together the judgements of several people in a way that prevents personal factors such as personality and rank having an influence.

Deterministic (in simulation) means that the inputs and assumptions are known and fixed c.f. Stochastic.

A *discrete distribution* is one for which there are gaps in the range of values the variable takes. Integer variables and classified data both give rise to discrete distributions.

A *dual value* (or shadow price) is the real worth of a resource to the decision-maker. In linear programming a dual value of a constraint is the increase in the objective function when 1 extra unit of that resource is available.

An *econometric model* is a set of equations which quantify and predict economic variables.

Estimation is the prediction of the values and population parameters from sample data.

Expected monetary value (EMV) is a decision-taking criterion calculated by weighting payoffs with their respective probabilities of occurrence. It is the average benefit or loss resulting from a decision or chance event if it were repeated many times.

Expected value of perfect information (EVPI) is the average additional benefit to accrue from a decision if the outcome of all chance events were known with certainty and the decision repeated many times.

Expected value of sample information (EVSI) is the average additional benefit to accrue from a decision if sample information (at 0 cost) were used and the decision repeated many times.

Exponential smoothing is a forecasting method based on forming a new smooth series by giving weights (of a particular type) to the original series.

Extrapolation is the use of a regression equation for predictions outside the range of conditions on which it was estimated. Typically this occurs when predictions are based on x values very different from those of the data set from which the coefficients were calculated.

The *feasible region* in linear programming is the set of all combinations of values of the decision variables that satisfy all the constraints.

A *fitted y value* is the y value of a point on a regression line (or curve) which has the same x value(s) as an actual data point.

A *flow chart* is a diagrammatic representation of the stages of a simulation.

Heteroscedasticity refers to the residuals in a regression analysis. When the residuals have different variances at different points of the line or curve they are heteroscedastic. If they have constant variance they are homoscedastic.

A *histogram* is a graph in which the horizontal axis carries the values of a data series and the vertical axis carries the frequencies with which those values occur.

Holt's method of forecasting is based on exponential smoothing. It can be applied to series with a trend.

An *index* is a measure summarizing the movement of some quantity over time. The best-known example is probably the cost of living index.

Infeasible LP problems are ones for which it is impossible to satisfy all constraints simultaneously.

Inference is the use of sample data to make predictions about the population from which the sample came.

The *intercept* of a straight line is the point at which it crosses the *y* axis. It is *a* in the equation of a straight line: $y = a + bx$.

Interquartile range is a measure of scatter. It is the difference between the highest and lowest numbers in what remains of a data series after the top 25% and bottom 25% have been removed.

An *isoprofit line* in a graphical LP solution is a line along which the profit objective is the same for all values of the decision variables. An *isocost line* refers to LPs in which the objective is to minimize cost.

A *Laspeyres index* is a weighted aggregate index where the weights are the quantities purchased in some base time period.

A *leading indicator* is a variable highly correlated with a variable to be forecast, whose movements are in advance of those in the forecast variable.

Least squares is the common criterion on which regression is based. The least squares line is the one for which the sum of squared residuals is a minimum.

Linear programming (LP) is a technique for solving decision problems characterised by being expressible algebraically in terms of a linear objective function and linear constraints.

Linear regression is regression in which the relationship between the variables is a straight line or (mathematically) of the form: $y = a + bx$.

Location means the general level of a data series. Arithmetic mean, median and mode are measures of location.

Maximin is a decision criterion by which the decision chosen is that whose worst outcome is the better than the worst outcome of any other decision.

Mean absolute deviation is a measure of scatter defined by:

$$\text{MAD} = \Sigma \, |x - \bar{x}| / n$$

Mean square error is a measure of forecasting accuracy. The differences between actual and forecast values are (1) squared (2) summed (3) averaged. A forecasting method should have a low MSE.

Median is a measure of location defined as the middle value of a data series. There are as many numbers in the series higher than the median as there are lower.

Mode is a measure of location defined as the most frequently occuring value in a data series.

Monte Carlo technique is a simulation technique which handles uncertain variables represented by probability distributions. The output of the simulation is also in the form of probability distributions.

Moving averages is a forecasting method based on forming a new smooth series by averaging groups of numbers in the original series.

Multi-collinearity occurs when two or more of the independent variables in a regression are correlated with one another. As a result their estimated coefficients are not reliable and should not be used for predicting.

Multiple regression is a method for finding the formula relating one variable (the dependent or y variable) to several others (the independent or x variables).

The *normal distribution* is a standard distribution deriving from the following situation. Repeated measurements are made of the same physical quantity. The measurements are subject to many small, independent, random, additive disturbances. The resulting distribution is normal.

An *objective function* in LP is an algebriac expression measuring the quantity to be optimised. Typically it is the equation representing profit.

An *observed distribution* is a collection of observed values of a variable. It is usually expressed as a histogram.

Outliers are extreme observations in a data series. Their values are very different from the great majority of the other observations.

A *Paasche index* is a weighted aggregate index in which the weights are the quantities purchased in the most recent time period.

Parameters are measurements which describe a population. They fix the context in which a variable varies.

Payoffs are the outcomes (usually monetary) when the end of a decision tree is reached. Each ending branch of a decision tree has a payoff.

The *Poisson distribution* is a standard distribution derived from the following situation. Isolated events occur in a continuum. A random sample of the continuum is taken. The Poisson distri-

bution gives the probability of a given number of events occuring in the sample. Typically the continuum is a period of
time.

A *population* is the set of all possible values of a variable.

Posterior probabilities. See *Bayes theorem*.

A *present value* is an amount of money received now which is equi-
valent to income received at points in the future. Present value
is a way of comparing cash received at different points in time.

Prior probabilities. See *Bayes theorem*.

Probability is a measure of the likelihood that an event will take place.
The probability of any event is a number between 0 (if it is
impossible) and 1 (if it is certain).

Random means in no particular order. A random sample is one in
which each member of the population has an equal chance of
being chosen.

Random number tables are lists of numbers selected by a random
process. The numbers are therefore in no particular order.

A *random walk* is a data series in which the difference between each
number and its predecessor is random.

Range is the difference between the highest and lowest values in a
data series.

Residuals in regression are the differences between the actual and
fitted *y* values. They are what is 'left over' when a line (or
curve) is fitted to a set of points.

Risk analysis is a financial simulation based on the Monte Carlo
technique.

Samples are selections from, or subsets of, populations.

The *sampling distribution of the mean* is the distribution of the
arithmetic means of a series of samples all of the same size.
Provided the sample size is greater than 30 the distribution is
normal whatever the distribution of individual elements of the
population.

Scatter means the variability or dispersion in a data series. Range,
mean absolute deviation, variance and standard deviation
are measures of scatter.

A *scatter diagram* is a graph in which paired observations of two
variables are plotted against each other.

Seasonality in a data series is a regularly repeating pattern of upward
and downward movements. The pattern takes a year or less to
repeat itself.

Sensitivity analysis is the repetition of a technique with the underlying assumptions varied to test the robustness of the results to changing circumstances.

Serial correlation refers to the residuals in a regression analysis. If there is a pattern in them by virtue of each residual being related to the previous one then they are serially correlated.

Shadow prices. See *Dual values*.

A *significance level* marks the accept/reject boundary in a significance test. If the sample evidence has a probability (assuming the truth of the hypothesis) lower than the significance level, the hypothesis is rejected; if higher, it is accepted.

Significance tests are means of deciding whether a particular piece of sample information is inconsistent with a supposition (or hypothesis) about the population.

A *simple aggregate index* is a simple index formed from the series which is itself the sum of several data series.

A *simple index* is formed when a data series is transformed so that it is based on 100.

Simple regression is a method for determining the formula relating one variable (the dependent or *y* variable) to another (the independent or *x* variable).

Simplex is a method (or algorithm) for solving LP problems.

Simulation is the modelling, physically or mathematically, of some system as an aid to planning and decision-making.

Skewness means non-symmetry. A histogram or distribution in which the observations are clustered around low values but with a few very high values is *right-skewed*; if the observations are clustered around very high values but with a few very low values, it is left-skewed.

Slack constraints in LP are ones for which all of the resource specified by the constraint is not used up at the optimal point. The amount by which the left-hand side is less than the right-hand side is the slack.

The *slope* of a straight line is its gradient, the ratio between the distance moved vertically and the distance moved horizontally. It is equal to b in the straight line equation $y = a + bx$.

A *spurious correlation* is one for which there is no real underlying relationship between the variables but, for some reason, the correlation coefficient is high.

Standard deviation is a measure of scatter defined by:

$$\text{Standard deviation} = \sqrt{\text{Variance}}$$

A *standard distribution* (or probability distribution or theoretical distribution) is a spread of values of a variable for which the probability of the variable taking any value has been calculated mathematically.

Standard error is the name often given to the standard deviation of a sample statistic when it has been estimated from one sample rather than measured from several samples.

Stationary series fluctuate about some constant level. They have no trend and constant variance over the length of the series.

Stochastic (in simulation) means that at least some of the inputs or assumptions are uncertain, being subject to probability distributions.

Surplus constraints in LP are 'greater than or equal to' constraints for which, at the optimal point, the specified minimum is exceeded. The amount by which the left-hand side exceeds the right-hand side is the surplus.

A *one-tailed test* is a significance test in which the hypothesis will be rejected by sample measures (for instance, the sample mean) falling only in one tail of the sampling distribution. In a *two-tailed test* the hypothesis will be rejected by sample falling in either tail.

Tight constraints in LP are ones in which the whole of the resource is used up at the optimal point. The left- and right-hand sides of the constraints are equal.

A *time series* is a series of measurements of a variable at regular time intervals. Time series forecasting methods are based solely on the historical record of the variable to be forecast.

Transformations of variables are used in regression analyses to convert a curved relationship into a linear one.

A *trend* in a data series is a consistent upward or downward movement.

Unbounded LP problems are ones whose formulation allows one or more variables to take infinite values.

Variables are measurable entities. The measurement varies each time an observation of it is made.

Variance is a measure of scatter defined by:

$$\text{variance} = \Sigma(x - \bar{x})^2/(n - 1)$$

Venn diagrams are diagrammatic ways of revising probabilities using Bayes Theorem.

A *weighted aggregate index* is a price index. It is formed from several price series which are weighted according to some criterion.

Index